D0741970

Quaternion Unified SuperStandard Theory (The QUeST) and Megaverse Octonion SuperStandard Theory (MOST)

4-Dimensional Biquaternion Universe Implies Standard Model and Unified SuperStandard Theory
8-Dimensional Bioctonion Megaverse Contains Biquaternion Universes with the Unified SuperStandard Theory
MOST Extends the Unified SuperStandard Theory
MOST Differentiates the Normal and Dark Sectors of Universes and the Megaverse
QUeST Accounts for Prportions of Dark and Normal Universe Mass-Energy
MOST Spinors May Help Explain Lack of Interactions with Dark Matter
Quantum Functionals Give Instantaneous Quantum Entanglement without Relativity Violation
Einstein Physical Reality Clarified and Extended
Quantum Functionals Implement Plato's Theory of Ideals
Quantum Functionals Implement Eastern Unity of the Cosmos

Stephen Blaha Ph. D.
Blaha Research

Pingree-Hill Publishing
MMXX

ISBN: 978-1-7328245-5-3

Rev. 00/00/01 January 6, 2020

To Margaret

Some Other Books by Stephen Blaha

All the Megaverse! Starships Exploring the Endless Universes of the Cosmos using the Baryonic Force (Blaha Research, Auburn, NH, 2014)

SuperCivilizations: Civilizations as Superorganisms (McMann-Fisher Publishing, Auburn, NH, 2010)

All the Universe! Faster Than Light Tachyon Quark Starships & Particle Accelerators with the LHC as a Prototype Starship Drive Scientific Edition (Pingree-Hill Publishing, Auburn, NH, 2011).

Cosmos Creation: The Unified SuperStandard Model, Volume 2, SECOND EDITION (Pingree Hill Publishing, Auburn, NH, 2018).

Immortal Eye: God Theory: Second Edition (Pingree Hill Publishing, Auburn, NH, 2018).

Unification of God Theory and Unified SuperStandard Model THIRD EDITION (Pingree Hill Publishing, Auburn, NH, 2018).

The Exact QED Calculation of the Fine Structure Constant Implies ALL 4D Universes have the Same Physics/Life Prospects (Pingree Hill Publishing, Auburn, NH, 2019).

Unified SuperStandard Theory and the SuperUniverse Model: The Foundation of Science (Pingree Hill Publishing, Auburn, NH, 2018).

Available on Amazon.com, bn.com Amazon.co.uk and other international web sites as well as at better bookstores (through Ingram Distributors).

CONTENTS

FIGURES and TABLES

INTRODUCTION

Previous derivations of the Unified SuperStandard Theory were based on Complex General Relativity and Quantum Field Theory. This book presents a deeper derivation that unites space-time coordinates and internal symmetry coordinates in a manner that does not violate No Go theorems of the 1960's. Our motivation is the close analogy between Lorentz subgroups and Standard model groups suggesting a possible common origin.

The result is a new basis for the Unified SuperStandard Theory. We call this new form the Quaternion Unified SuperStandard Theory (QUeST). It includes the Unified SuperStandard Theory and the Standard Model.

Assuming a higher dimension Megaverse also exists within which our universe resides, we develop a more comprehensive theory that includes the Unified SuperStandard Theory. This theory provides a clear reason for the lack of interactions between normal and Dark matter. The new theory is the Megaverse Octonion SuperStandard Theory (MOST). It is of interest because it extends and clarifies the symmetries of QUeST (our universe); because it explains the parallel between the subgroups of the Lorentz group and the Standard Model symmetry groups, and because it provides a clear reason for the lack of interactions between normal matter and Dark matter.

Quantum functionals are also described in some detail. We show that the Einstein, Podolsky, and Rosen definition of physical reality needs to be extended. We also show that our quantum functionals, because they are all located at a point without any separation distance, support instantaneous quantum action-at-a-distance without spookiness.

1. Seeking a Deeper Basis of Physical Reality

We have strived in our earlier books to delve more deeply into the fundamental basis of the Cosmos. The hard won experience of past generations has brought us to the brink of understanding the nature of physical reality. Our recent development of the Unified SuperStandard Theory and the Megaverse Model based on the axioms of Fig. 1-A.1 gives a clear consistent understanding of elementary particle, gravitation, and the overall structure of the universe. It encompasses what we know of reality and adds logically justifiable new features, which we hope will eventually be found experimentally.

Now we consider a possibly deeper view of Physical Reality based on the close analogy of the subgroups of the Complex Lorentz group (U(1), SU(2), and SU(3)) with the known symmetry groups of the Standard Model as well as our Unified SuperStandard Theory. The similarity raises the hope of a common deeper origin.

We pursued this possibility to simplify the basis of the theory but more importantly to conclusively nail down the symmetry structure of the ultimate theory. *We found a common parent for the siblings: the Lorentz group of space-time, and the internal symmetries of the theory.* Although the parentage is the same, the siblings are disjoint, and thus violations of the "NoGo" theorems of the 1960's are avoided.

We now have a theory that has a deeper basis, and yet gives the features of the Unified SuperStandard Theory. We call this theory Quaternion Unified SuperStandard Theory (QUeST). Chapters 5 and 6 described the features of the deepened theory. Chapter 6 has the new, somewhat revised, set of axioms that are explained in some detail.

A further study of the possibilities of the unified SuperStandard theory in the Megaverse (the space in which our universe may reside) led us to a yet more comprehensive theory that includes QUeST but adds more symmetry in a manner that seems physically reasonable. We call this greater, all encompassing, theory the Megaverse Octonion SuperStandard Theory (MOST).

The symmetry groups of MOST are a superset of those of QUeST, which in turn is a superset of the Unified SuperStandard Theory of our books Blaha (2019g) and (2018e) as well as earlier books. The symmetry of MOST is quite remarkable in its capture of *all* the symmetry groups of the Unified SuperStandard Theory, and its addition of new internal symmetries. MOST also shows why Dark particles do not interact with normal particles. It is due to the nature of MOST fermion spinors and vector boson spins.

The major consequences of the deeper justification are:

1. *Space-time structure and group symmetries emerge from the same formalism* while they are not combined, but are independent, unlike abortive attempts at unification like the SU(6) of the 1960's.

2. The structure of the Megaverse becomes clearer particularly in regard to its relation to the structure of the universes within it. The Megaverse in MOST has 8 complex dimensions.

3. Group symmetries are clarified due to their interrelation within a larger framework. Megaverse internal symmetries are shown to be a superset of Unified Standard Theory internal symmetries

4. Particle symmetries and interactions, and the spectrum of particles, are determined by a deeper fundamental structure of space and time.

5. The major features of the Unified SuperStandard Theory follow directly from the space and time structure of the deeper level of reality.

6. Some axioms of the Unified SuperStandard Theory are placed on a deeper, firmer foundation. See chapters 6 and 13.

1.1 Previous Basis for the Unified SuperStandard Theory

In previous books we traced the origin of Chemistry and Biology ultimately to Physics. We further traced the basis of Physics to the Physics of Elementary Particles and of General Relativity. We based the Unified SuperStandard Theory on Complex General Relativity and Quantum Field Theory.

We choose to start with Complex General Relativity because it leads to Complex Special Relativity which subsequently leads to the general form of the spectrum of spin ½ fermions as a set of four types of fermions: charged leptons, neutral leptons, up-type quarks and down-type quarks.[1]

Complex General Relativity is one of the parents in the Foundation of Physics. The other parent is Quantum Field Theory.[2]

One might ask why this choice of parentage was made. We argue the key factors that lead to the choice are motion and creation/destruction. Motion requires coordinates for its specification. Complex General Relativity offers a most general formulation leading to coordinates and the description of motion. Needless to say, motion requires entities that move. Particles are perhaps the simplest entities. The acts of creation and destruction seem to be a requirement of particles to avoid a static universe. The simplest method of implementing creation and destruction is Quantum Field Theory. It avoids the alternatives of interacting blobs of matter and of more complex forms of entities such as strings. Thus we have a simple rationale for the Foundation of Physics.

This Foundation led to the known Standard Model in a fairly direct way and beyond that to our Unified SuperStandard Theory.

1.2 A Yet Deeper Basis for the Unified SuperStandard Theory

The combination of Complex General Relativity and the particle nature of Quantum Field Theory lead us to consider the possibility that a very deep connection exists between the coordinates of General Relativity and the coordinates of the fundamental representations of the internal symmetry groups of particle physics. To that

---

[1] One cannot have a Complex Special Relativity without embedding it in a Complex General Relativity.

[2] We do not accept a deeper foundation for Fundamental Physics, such as Emergent Physics, due to the total absence of experimental evidence for a deeper level.

end we consider a unified formulation of physical space that unites space-time coordinates with fundamental representation coordinates of the particle internal symmetry groups. Chapters 5 and 6 present the formulation for our universe embodying this approach. It successfully yields space-time as we know it, and the known and new internal symmetry groups of the Unified SuperStandard Theory

Chapter 13, and subsequent chapters, takes the approach further based on the assumption that our universe resides in a larger space, the Megaverse. We find the extended theory that we developed has 8 complex dimensions. It directly yields the known internal symmetries of the Standard Model, the symmetries of the Unified SuperStandard Theory, and additional symmetries that make physical sense. The tightness of the internal symmetry structure it develops is very encouraging. We do not make *ad hoc* assumptions about the choice of symmetry groups as is so often done. The development of the theory from the deeper basis naturally leads to the symmetry groups.

The deeper view that it produces of the reason for the non-interaction of normal matter and Dark matter is compelling. See chapter 13 and 14 for details..

Appendix 1-A. The Previous Basis of the Unified SuperStandard Theory

It seems worthwhile to review the previous basis of the Unified SuperStandard Theory because the new deeper basis builds on it. This appendix is an extract from Blaha (2019g).

1-A.1 A Foundation in Complex General Relativity

We now turn to follow the path that ultimately leads to the SuperStandard Theory. Unlike other attempts at a fundamental theory we do not posit groups for interactions but rather provide strong arguments (derivations) for the choice of the Standard Model group and the SuperStandard Theory group.

We assume that the fundamental theory of space-time coordinates is Complex General Relativity. It is General Relativity extended to complex-valued coordinates with a complex-valued energy-momentum tensor. Complex General Relativity can be factored into a U(4) group that rotates complex coordinates and a residual Complex General Relativity factor. We call the U(4) group the Coordinates Species group for reasons given later.

From Complex General Relativity we proceed to consider the flat space-time case with Complex Special Relativity. Complex Special Relativity is described by the Complex Lorentz Group which has the subgroups SU(2), U(1), SU(3), and additional SU(2) and U(1) subgroups.

After showing a map from coordinate subgroup symmetries to elementary particle group symmetries we find the Coordinates Species group maps to a U(4) Particle Species group.

1-A.2 A Foundation in Quantum Field Theory

Particles, and particle symmetries, to which we have been alluding above, emerge from Quantum Field Theory—the other Foundation of Physics. Why do we view Quantum Field Theory as a fundamental foundation? Of all the forms a dynamics theory might take, Quantum Field Theory is the simplest form that supports the creation and annihilation of matter—a fundamental attribute of matter as we know it. Blob creation and annihilation would take us into complexity as would realistic string creation and annihilation.

Quantum Field Theory offers a simplicity that is easily seen in the representation of creation and annihilation using Feynman diagrams.

Thus we opt for Quantum Field Theory and find it, and Complex General Relativity, sufficient to describe all known features of elementary particles and their combinations into more complex forms of matter and energy. As we saw in Blaha (2019g) and (2018e) and our earlier books the purported problems of Quantum Field Theory are easily curable.

A direct benefit of Quantum Field Theory is the appearance of particle number operators which lead to the Generation Group and the Layer Groups

1-A.3 Axioms of the Unified SuperStandard Theory

Complex General Relativity and Quantum Field Theory lead to the Unified SuperStandard Theory. The fundamental axioms that specify the basis of the derivation of the Model are listed in Fig. 1-A.1.[3] Their detailed implications are:

1. Each space-time symmetry subgroup of the Complex Lorentz Group has a corresponding particle interaction symmetry group. The particle symmetry groups combine in a direct product.

   The specific subgroups of the Complex Lorentz groups with corresponding particle interaction symmetry groups is only well-defined if we further require that they correspond to the distinction between space and time. Thus the SU(3)

[3] Chapters 6 and 13 provide a revised deeper set of axioms for our universe and the Megaverse respectively.

subgroup emerges from the 3 × 3 space part of Complex Lorentz Group elements. The pair of SU(2) ⊗ U(1) subgroups emerge from the "boost" parts of Complex Lorentz Group elements. Consequently the corresponding enlarged Standard Model particle interaction symmetry is

$$SU(2)\otimes U(1)\otimes SU(3)\otimes SU(2)\otimes U(1)$$

The minimal group in which the above direct product is a subgroup is SU(7) due to the Dark SU(2)⊗U(1) factor.

2. Quantum Field Theory supports fundamental particles that form a countable set. Each particle number operator is a generator of a particle interaction group.

   All matter and energy is composed of discrete particles.

3. All quantum field theory calculations are finite.

   Perturbation theory calculations have divergences in conventional quantum field theory that require renormalization. Axiom 3 below leads to the author's Two Tier formulation[4] where there are no divergences—eliminating the need for renormalization to eliminate divergences.

4. The Quantum Field Theory of particles can be defined in any curved space-time.

   In certain curved space-times conventional second quantization leads to ambiguities in the definition of particle states. Axiom 4 below leads to the author's generalized second quantization procedure,[5] called Pseudoquantization,

---

[4] See Blaha (2005a).

[5] S. Blaha, Phys. Rev. D**17**, 994 (1978) and references therein to earlier papers by the author such as Phys. Rev. D**10**, 4268 (1974) and Il Nuovo Cimento **49A**, 35 (1979) and **49A**, 113 (1979).

which eliminates these ambiguities. It also supports a canonical lagrangian formulation of higher derivative field theories.[6]

5. Each particle wave function has a quantum functional[7] defining the particle state in a space without a distance measure.

   There is a set of functionals (called monads or cores) with an element for each particle in the universe.[8] The quantum entanglement of particles at a distance can be instantaneous because the functionals (which embody the state of each particle) exist in a space without a distance measure. In a sense this feature embodies the unity of creation.

The five axioms imply the detailed list of axioms in Blaha (2019g) and (2018e).The axioms are revised for QUeST (chapter 6) and also for MOST (chapter 13).

---

[6] S. Blaha, Phys. Rev. **D10**, 4268 (1974) and **D11**, 2921 (1975) and references therein.
[7] See Blaha (2019g) and (2018e) for a detailed discussion. Our approach eliminates the issues of "spookiness" and instantaneous action at a distance that clouds quantum entanglement.
[8] See Blaha (2019g) and (2018e).

AXIOMS

1. Each space-time symmetry group of the Complex Lorentz Group has a corresponding particle interaction symmetry group. The particle symmetry groups combine in a direct product.

2. Quantum Field Theory supports fundamental particles that form a countable set. Each particle number operator is a generator in a particle interaction group.

3. All quantum field theory calculations are finite.

4. The Quantum Field Theory of particles can be defined in any curved space-time.

5. Each particle wave function has a functional defining the particle state in a space without a distance measure.

Figure 1-A.1 The axioms of the Unified SuperStandard Theory.

2. Quaternion Unified SuperStandard Theory (QUeST) Formulation Compared to Unified SuperStandard Theory

This chapter lists some of the highlights of the Unified SuperStandard Theory (presented in Blaha (2018e) and (2019g)) and briefly identifies the differences from our new Quaternion Unified SuperStandard Theory (QUeST) formulation presented in this book for the first time. QUeST gives a deeper foundation. It has the benefit of corresponding to the known features of the Standard Model and most of the features of the Unified SuperStandard Theory. *It solidifies the symmetry structure of the theory into a concrete pattern and thus avoids the use of conjectural ansätze to determine elementary particle symmetries.*

1. The number of spatial dimensions was determined to be the number of generators in the primary set of interactions of the space. In the case of an *empty* universe the primary set of interactions is the U(2) qubit transformations group. The number of U(2) generators is four and thus the dimension of space is 4 complex dimensions. Also and more importantly, considerations of Asynchronous Logic, and the requirement that physical processes must be able to proceed in parallel, require the number of spatial dimensions to be four. The book justifies four complex space-time dimensions with a Lorentz metric yielding Complex Lorentz group symmetry.

Change: The dimensionality is set by the quaternion foundation..

2. Boosts of the Complex Lorentz group transform a Dirac-like equation with a Landauer mass into four different forms (called species). Each form maps to a type of fermion: neutral leptons (neutrinos), charged leptons, up-type quarks, and down-type quarks. Neutral leptons and down-type quarks are tachyons. Some evidence exists for tachyonic neutrinos. Complex Lorentz boosts lead to the Complex Lorentz group

factorization: SU(2)⊗U(1)⊗SU(3)⊗SU(2)⊗U(1). We map SU(2)⊗U(1)⊗SU(3) to fermion particle functional space to obtain the internal symmetry group for ElectroWeak and Strong Interactions: SU(2)⊗U(1)⊗SU(3). The remaining factors SU(2)⊗U(1) we map to the internal symmetry group for Dark Matter, which we take to be the Dark ElectroWeak Interaction (unconnected to normal matter interactions).

> Change: We find an additional SU(3) Dark Strong Interaction group giving a SU(2)⊗U(1)⊗SU(3)⊗SU(2)⊗U(1)⊗SU(3) symmetry.

3. Parity Violation as seen in the Weak Interactions follows directly from the forms of the four types of fermions predicted by the Complex Lorentz Group.

> Change: None

4. The existence of four conserved (and partially conserved) quantum numbers such as baryon number and lepton number indicates that there is a U(4) group whose 4 representation causes each species to have four generations—three of the generations are known. We suggest that a fourth generation of much higher mass fermions exist.

> Change: Normal matter has a U(4) Generation group and Dark matter has a separate U(4) Dark Generation group.

5. In each generation there are four partially conserved quantum numbers. Thus we find that there is another U(4) group (called a Layer group) for each generation yielding the combined Layer groups $[U(4)]^4$. The 4 representation of each U(4) results in a fermion spectrum of four layers of four generations or 192 fermions in all. We see only one layer at present. The additional three layers of fermions remain to be found at much higher masses. The symmetry group of the Unified SuperStandard Theory is

$$[SU(2)\otimes U(1)\otimes SU(3)\otimes SU(2)\otimes U(1)\otimes U(4)\otimes U(4)]^4\otimes U(4)$$

where the last factor is for the broken Species group, which follows from Complex General Relativity.

Change: Changed symmetry. Additional Dark Strong SU(3) group and separate Generation and Layer groups for the Dark sector to avoid generating normal particle and Dark particle interactions. The result:

$$[U(1)\otimes SU(2)\otimes SU(3)\otimes U(1)\otimes SU(2)\otimes SU(3)\otimes U(4)^4]^4\otimes U(4)$$

6. Assuming all particles are massless at the Big Bang, and all particle types have an equal proportion of the total mass-energy then, we find that the 192 fermions and 192 vector bosons yield a Dark Matter percentage of 83.33% (experimentally the estimates are 84.5% and 81.5%). The proportion of Dark Mass-Energy is found to be 91% of the universe's mass-energy. Experimentally the proportion has been estimated to be 95%. These results agree well with experiment. See chapter 14 of Blaha (2019g) and (2018e) for details.

Change: Now 128 Dark fermions and 128 "normal" fermions (of which 24 are known) totaling 256 fermions. Now 340 Dark vector bosons and 12 known vector bosons totaling to 352 vector bosons. As a result we find the percent of Dark matter is 90.6%, and the percent of Dark Mass-Energy is 96.6%., which are similar to the known data values. See Figs. 2.1 and 2.4.

7. The instantaneous quantum effects between space-like separated parts of a quantum state ('spookiness') is taken to be a feature of fundamental importance. The only sensible way to implement this feature in quantum theory is to assume that the wave function of every particle is the inner product of a particle functional and a wave (fourier) coordinate expansion. Particle functionals exist in a space with no distance measure. The space of coordinate expansions also has no distance measure. Other functionals in a state (and their implicit coordinate fourier expansions) change *instantaneously* when one of the functionals comprising a state changes since coordinate space distance is irrelevant.

Change: None

8. Fermion particle functionals are called *Qubes*. They exist 'within' every fermion. They have a mass that we take to be the Landauer mass—the minimal energy of a qubit. Boson particle functionals are called *Qubas*. They are assumed to be massless in the absence of all interactions to preserve free vector boson and spin 2 boson gauge symmetry. Free Higgs particles are assumed to be massless for consistency.

Change: None

9. To have a completely finite theory with no infinities (including no fermion triangle infinities) we introduced Two-Tier Coordinates that replaced normal point like coordinates with a type of 'fuzzy' coordinates.

$$X^\mu = x^\mu + iY^\mu(x)/M_c^2.$$

Change: None

10. Since the Unified SuperStandard Theory lagrangian would require higher order derivatives to account for quark confinement (linear potential terms) and for MoND-like deviations from conventional gravity, and since such terms would be outside a canonical lagrangian formulation, we introduced two fields for each particle (fermions and bosons) in a formulation we call Pseudoquantum Theory. Pseudoquantum theory enables a canonical lagrangian formulation. It has other advantages such as a clean separation of vacuum expectation values from quantum fields for Higgs particles. It also supports second quantization in arbitrary coordinate systems while maintaining the same particle interpretation of states in all coordinate systems.

Change: None

11. The book also describes Higgs symmetry breaking and the use of the Faddeev-Popov Mechanism in detail for the theory.

Change: None

12. Since a Complex Special Relativity requires a Complex General Relativity we considered Complex General Relativity and showed that it could be 'factored' into General Relativity and a new U(4) group that we called the Species group. Since Complex General Relativity must support interactions with all types of matter we specified a Species group interaction with all matter. Further, we assumed that the Species vector bosons acquired masses through the Higgs Mechanism The Higgs Mechanism caused Species group contributions to each fermion mass. Such a mass term would require each fermion particle mass to be both inertial *and* gravitational *solving the mystery of the equality of inertial and gravitational mass.*

Change: None

13. We showed that the implicitly higher derivative Riemann-Christoffel curvature tensor for all interactions leads to new interactions beyond The Standard Model. In addition to yielding quark confinement and MoND-like modifications of gravity, it may help understand the missing nucleon spin issue, discrepancies in proton radius measurements, vector meson dominance (VDM), and so on.

Change: Addition of Dark SU(3) terms and Dark Generation and Layer groups to the Riemann-Christoffel tensor for each layer.

14. We defined an Interaction Rotations group that caused rotations among all the vector boson interactions of The Unified SuperStandard Theory. We found that rotations that respected Superselection rules such as the Charge Superselection rule could have physical significance. One example is ElectroWeak Theory which is an application of Interaction Rotation transformations.

Change: The Interaction Rotations group is now not compelling since it is not in QUeST.

15. Since the number of fundamental fermions (192) and fundamental vector interaction bosons (192) is equal we considered Supersymmetric-like features of the Unified SuperStandard Theory.

Change: The number of fermions and vector bosons is changed. (Item 6.) SuperSymmetry not indicated.

16. The discovery of two new particles that do not appear to be within the framework of The Standard Model, as it is currently known, raises the possibility that they may be within the expanded fermion spectrum in The Unified SuperStandard Theory. Towards that end we present a *preliminary* assignment of the locations of the new fermions within the spectrum of the Unified SuperStandard Theory.

Change: The additional Dark Strong SU(3) group implies Dark quarks are triplets.

17. We showed that the coupling constants of the Standard Model including the Fine Structure Constant of QED are determined by eigenvalue functions. As a result they differ and in a way that appears contrary to the hypotheses of Grand Unified Theories (GUTs) even taking account of running coupling constant considerations.

Change: None

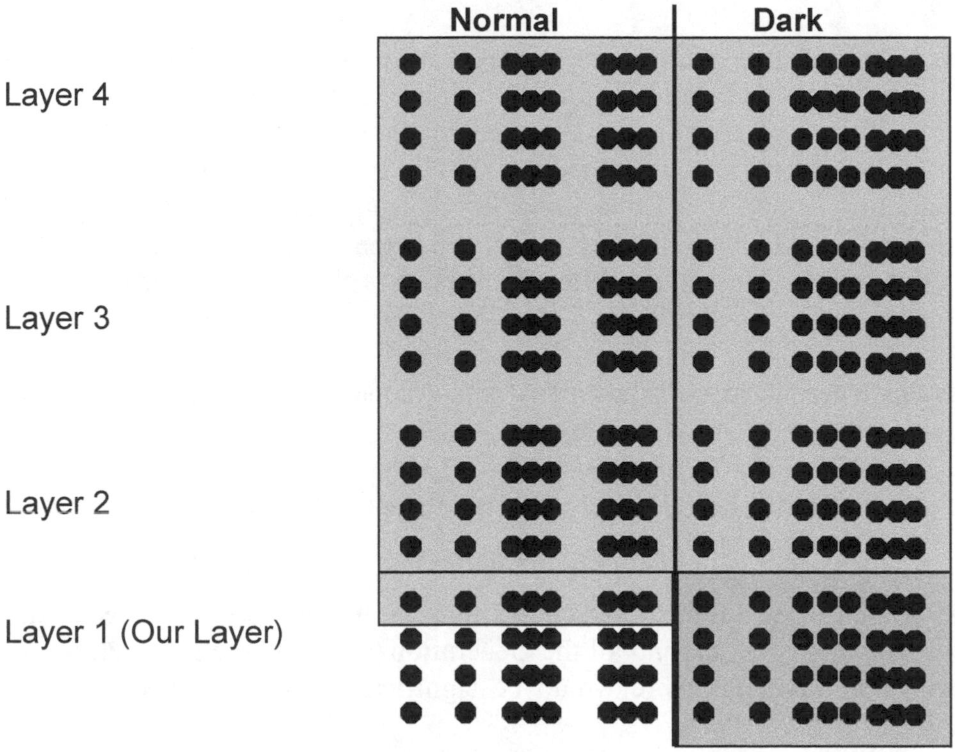

Figure 2.1 The Quaternion Unified SuperStandard Theory (QUeST) fundamental fermion spectrum consists of four layers of four generations of fermions. Current Dark matter parts of the periodic table are grayed. Light parts are the known fermions with an additional, as yet not found, 4th generation shown. The Normal and Dark matter parts are of similar form.

Implicit in our discussion of particles are the assumptions:

1. Fermion masses significantly increase from layer to layer. (Otherwise their particles would have been already found.)

2. Each layer has their own set of Standard Model interactions: Electromagnetic, Weak, and Strong, and their Dark Matter equivalents. The "electric" charge and internal quantum numbers of fermions are different from layer to layer. The interactions in the higher layers differ from those of the known layer by having larger vector meson masses.

3. We assume corresponding Standard Model interaction strengths (couplings) increase as one goes from layer to layer.

4. We assume Standard Model Interaction strengths are greater than the Generation group strength in each layer.

5. We assume all Layer groups interaction strengths are less than all the Standard Model interaction strengths and all the Generation group interaction strengths of all the layers. Otherwise there would be significant mixing between fermion layers.

The set of interactions (minus the Layer groups, the Species group and the Interactions Rotation group of the Quaternion Unified Super Standard Theory) is presented in Figs. 2.2 and 2.4.

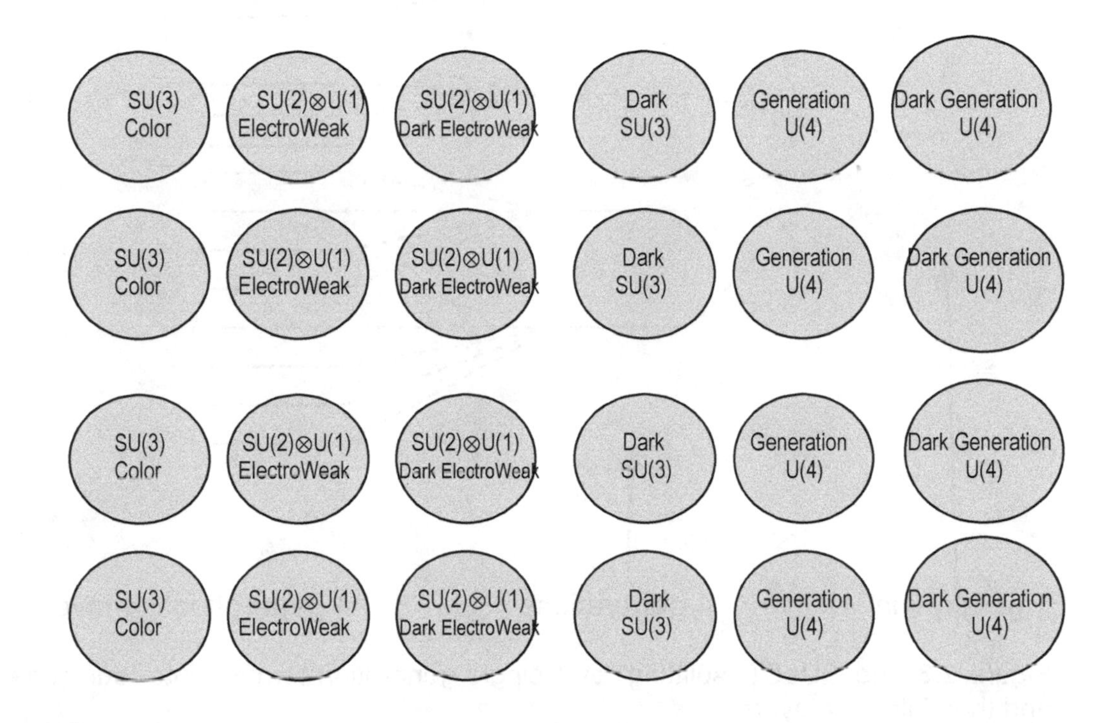

Figure 2.2. The QUeST set of four layers of internal symmetry groups corresponding to four generations in four layers of spin ½ fermions and the four layers of vector bosons. In addition there are the Normal and Dark Layer groups, the Species group and the Interaction Rotations group Θ that are *not* displayed.

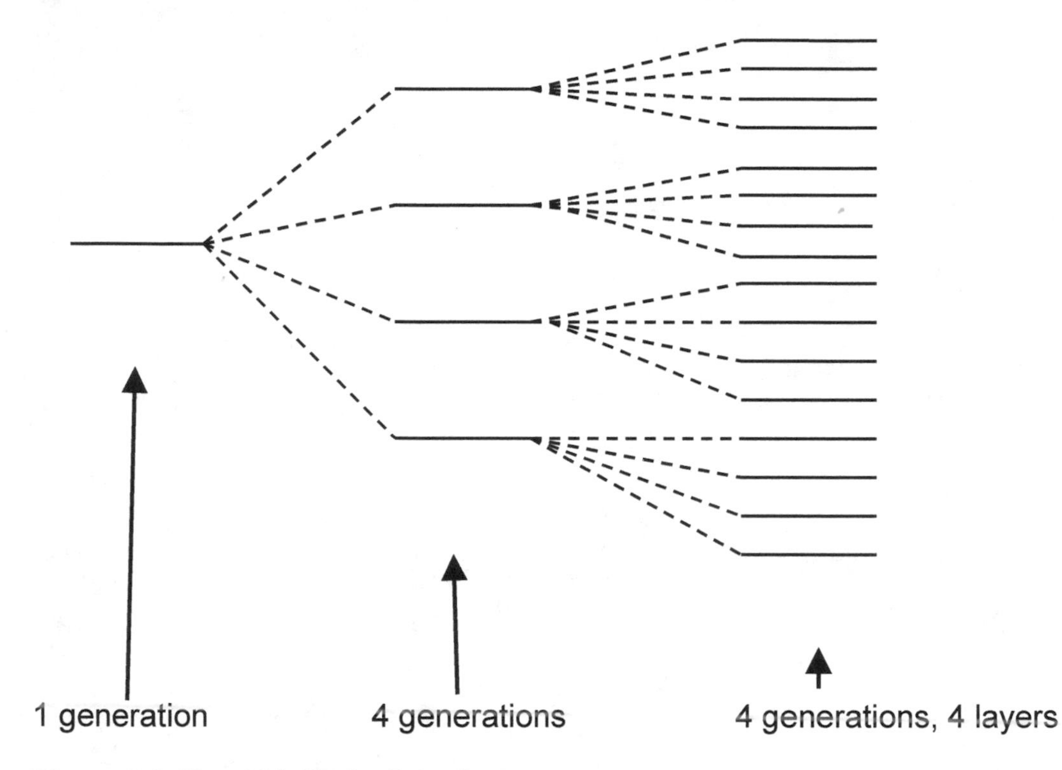

Figure 2.3 The QUeST "splitting" of a single generation fermion into four generations and then into four layers.

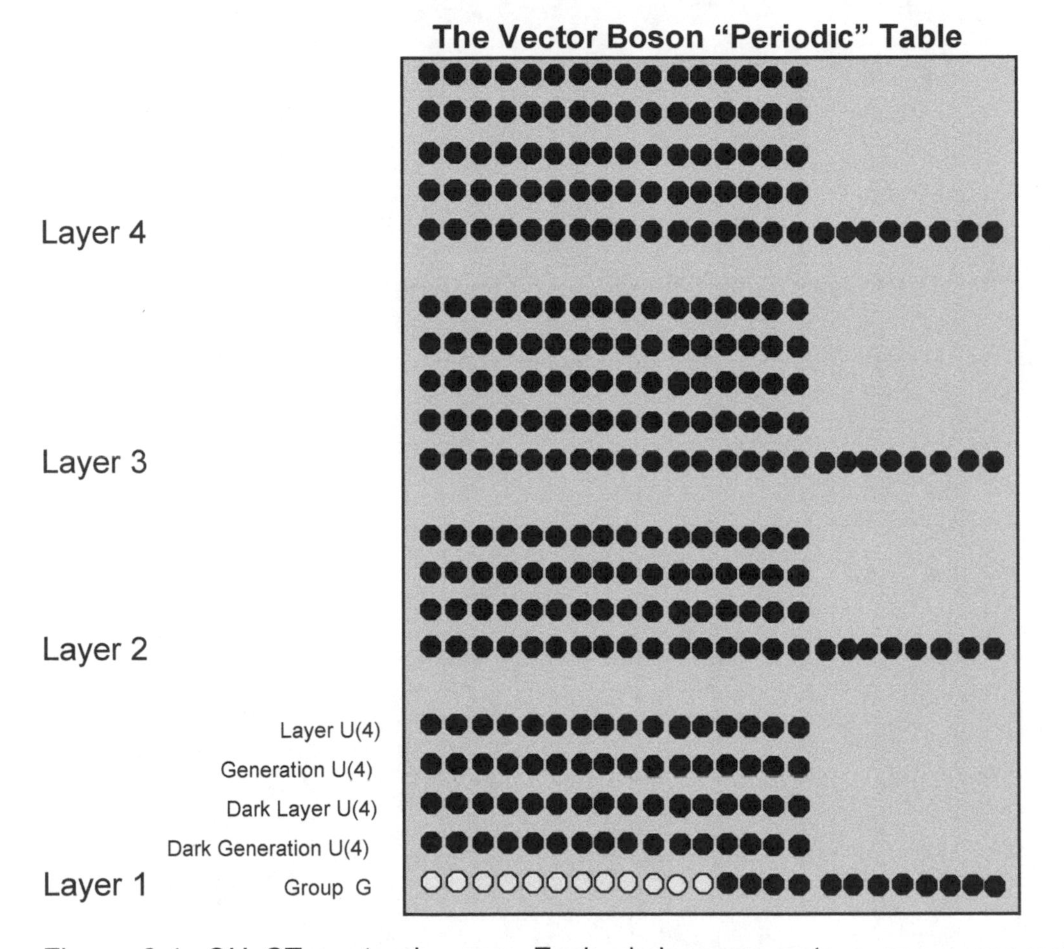

Figure 2.4. QUeST vector bosons. Each circle represents a group generator. The known vector bosons are in the lowest row with a white interior. Yet to be found vector bosons are solid black. The Layer groups are distributed by layer symbolically although they each straddle all four layers. G is $SU(3)\otimes SU(2)\otimes U(1)\otimes SU_D(2)\otimes U_D(1)\otimes SU_D(3)$. The list of groups for the higher three levels is the same as those of the first layer. There are 352 vector bosons.

Appendix 2-A. Gauge Fields Based on Particle Numbers

In this Appendix we show thc origin of the Generation and Layer groups in particle number operators. Particle interactions followed directly in the Unified SuperStandard Theory by analogy with Complex General Relativity subgroups yielding

$$SU(2)\otimes U(1)\otimes SU(3)\otimes SU(2)\otimes U(1)\otimes SU(3) \tag{2-A.1}$$

where the latter three factors are the Dark interactions.

They have a SU(10) covering group that contains this direct product of groups. The groups in eq. 2-A.1 are particle interaction groups in the Unified SuperStandard Theory.

Unlike other attempts to develop a formulation of the Standard Model (or generalizations) the Unified SuperStandard Theory was originally directly based on a theory foundation consisting of Complex General Relativity and Quantum Field Theory. Later we will show a deeper basis in Quaternions and Octonions.

To those who might prefer to base a theory on real General Relativity we note that proofs in Quantum Field Theory *require* the Complex Lorentz Group.[9] Thus the Complex Lorentz group is unavoidable for a properly (and rigorously) formulated Quantum Field Theory. Since the formulation of the Complex Lorentz Group in flat space-time can only be as the limit of Complex General Relativity, the choice of a foundation of Complex General Relativity is required.

Since particles are countable, and thus have discrete particle numbers, Quantum Field Theory brings particle numbers, and particle number laws such as particle conservation laws, into consideration.

---

[9] Streater (2000).

Blaha (2019e) and earlier books showed that Complex Lorentz boosts generate four types of fermion particles that we call *particle species*. We map these four species to charged leptons (such as electrons), neutral leptons (such as neutrinos), up-type quarks (such as the u quark), and down-type quarks (such as the d quark).

11-A.1 Basis of the Generation Group

We define two particle number operators for normal up-quark particles and down-quark particles, B_{uq} and B_{dq}. Similarly we define two particle number operators for normal species "e" (electron) particles and species "v" particles, B_e and B_v. Similarly we define Dark matter equivalents:[10] B_{De}, B_{Dv}, B_{Duq}, and B_{Ddq}.

In the absence of interactions these fermion particle number operators are conserved. Each set are "diagonal" operators within a U(4) group. Thus we have a normal U(4) Generation Group and a Dark U(4) Generation group.

On this basis we find there are four generations of each species in the normal and in the Dark matter sectors. One generation of normal fermions with large masses have not as yet been found.

The gauge vector bosons of the Generation Group also have large masses. If the conservation of the fermion particle numbers is broken then we view it as a consequence of Generation Group symmetry breaking.

2-A.2 Basis of the Layer Group

The set of particle number operators can be further refined if we take account of the fourfold fermion generations. To further refine the set of particle number operators we temporarily neglect all interactions that would violate conservation laws for the set.

We therefore subdivide the above particle number set into four particle numbers per generation. For the ith generation we define

L_{ie} – The "e" species particle number for the ith generation

[10] By analogy, we assume that there are four species of Dark matter: charged Dark leptons, neutral Dark leptons, Dark up-type quarks, and Dark down-type quarks. Thus we are led to the Dark particle numbers: Dark Baryon Numbers, and Dark Lepton Numbers shown above.

$L_{i\nu}$ – The ν species particle number for the ith generation
L_{iuq} – The up-quark species particle number for the ith generation
L_{idq} – The down-quark species particle number for the ith generation

L_{iDe} – The Dark "e" species particle number for the ith generation
$L_{iD\nu}$ – The Dark ν species particle number for the ith generation
L_{iDuq} – The Dark up-quark species particle number for the ith generation
L_{iDdq} – Dark down-quark species particle number for the ith generation

for each generation i = 1, 2, 3, 4. Individual fermions have positive $L_{ia} = +1$ values and anti-fermions have negative $L_{ia} = -1$ values for species a = 1, 2, 3, 4 (with the three color subspecies of quarks treated as part of one species.)

At this point we have four particle number operators for each generation. We define a group framework for each set of particle numbers. The simplest way is to assume that each generation consists of four layers with the particles in each generation in a U(4) fundamental representation.[11] Then each generation has a U(4) Layer group with the generation's four number operators (above) as its diagonal operators. We call this group the Layer Group of the ith generation L_{ia}. With four generations we obtain four U(4) Layer groups for normal matter. In addition there are four U(4) Dark Layer groups. See Fig. 2.4.

The consequence of this expansion of particle numbers and groups is that the set of fermions increases fourfold. We now have four layers, with each having four generations, Experimentally we know of three generations of fermions—the lowest generations of the lowest level. The remaining generation and three levels of fermions is of much higher mass and yet to be found.

See Blaha (2019g) and (2018e) for a detailed discussion of the Layer Groups. We note in passing that the symmetries of these number operators are badly broken. Yet the underlying group structure remains.

[11] See Fig. 2.3 for a depiction of the "splitting" of fermions: first into generations, then into layers.

3. The Bifurcated Universe

The content of our universe appears to be divided into two parts: a "normal" part that we see directly, and a Dark part that we only see indirectly through its gravitational effects. Thus we see a *bifurcated universe*.

The Dark part dominates both in its matter and energy. Attempts to find interactions between the Dark and normal sectors have hitherto failed. Thus the Dark and normal parts have extremely weak interactions or no interactions between them except gravitation. We suggest a deeper reason for the absence of interactions based on fermion spinor considerations in chapter 14.

The Unified SuperStandard Theory and the Quaternion Unified SuperStandard Theory (QUeST)y have a similar set of interactions and a similar particle spectrum with one exception: QUeST has an additional Dark Strong SU(3) interactions for Dark quarks and thus Dark quarks are triplets. (Chapters 13 and 14 describe QUeST interactions in detail.)

Separating the Normal and Dark sectors we find the gauge vector boson interactions are:

	Normal	**Dark**
	$SU(2)\otimes U(1)\otimes SU(3)$	Dark $SU(2)\otimes U(1)\otimes SU(3)$
Generation Groups	$U(4)^4$	Dark $U(4)^4$
Layer Groups	$U(4)^4$	Dark $U(4)^4$

We found we had to separate the Generation groups and Layer groups for the Normal and Dark sectors to preclude interactions between Normal and Dark energy and matter. The separated vector bosons are depicted in Fig. 3.1.

The separated Normal[12] and Dark fundamental fermions in QUeST are:

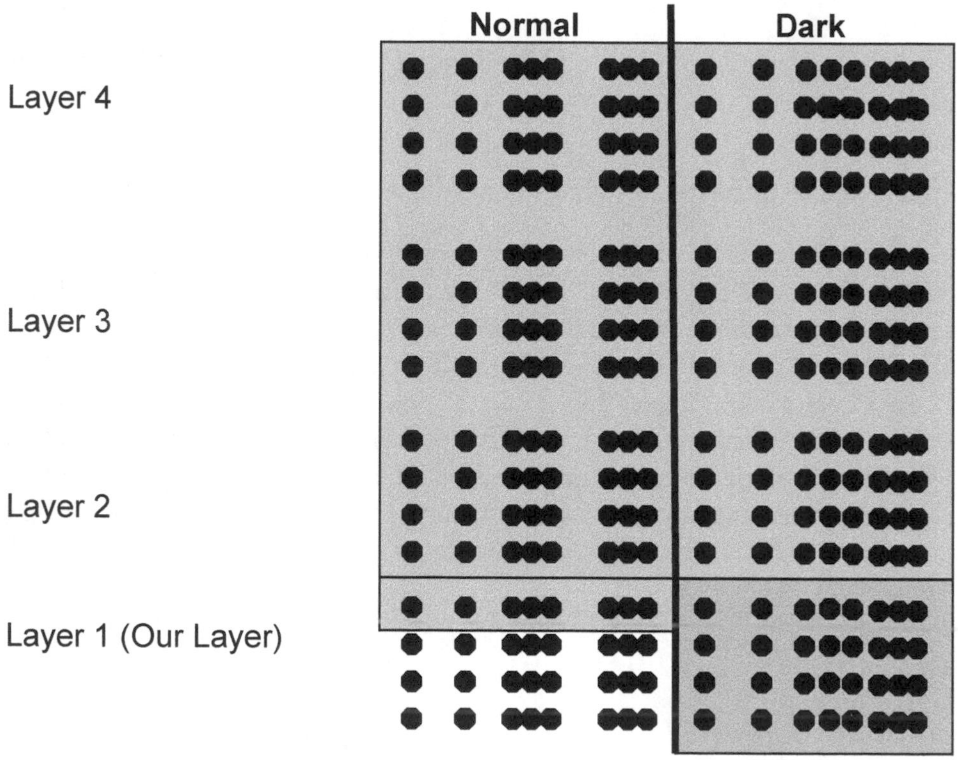

Normal Dark

Layer 4

Layer 3

Layer 2

Layer 1 (Our Layer)

Figure 3.1 The Quaternion Unified SuperStandard Theory (QUeST) fundamental fermions. Currently unknown new fermions are in grey boxes.

Note the Normal and Dark fermion spectrums have the same form.

[12] We call the known generations of the first level and the levels above it Normal. The fermions to the right in Fig. 3.1 are called Dark. Similarly the Normal and Dark gauge fields and interactions are separated as on the preceding page.

4. QUeST Riemann-Christoffel Curvature Tensor

In this chapter we will define the QUeST extension of the Riemann-Christoffel Tensor of the Unified SuperStandard Theory. The new part will consist of the addition of Dark SU(3) Strong interactions. Its addition will generate a confining Dark Strong interaction term for Dark quark confinement with an r potential—in a manner analogous to the normal quark confinement mechanism. The Dark Generation groups and Dark Layer groups terms are also explicitly shown.

In view of our goal of defining a unified theory of elementary particles and General Relativity we begin by defining a Riemann-Christoffel curvature tensor which we will use to construct a lagrangian for the theory.

4.1 The Covariant Derivative

The covariant derivative[13] which appears in fermion and gravitation equations uses the vector boson 7-vector:

$$ {}^{a}\mathbf{A}_{I}^{\mu}(x) = ({}^{a}g_{1}{}^{a}\mathbf{A}_{SU(3)}{}^{\mu}(x_{C}), {}^{a}g_{2}{}^{a}\mathbf{W}^{\mu}(x), {}^{a}g_{3}{}^{a}\mathbf{A}_{E}{}^{\mu}(x), {}^{a}g_{4}{}^{a}\mathbf{W}_{D}{}^{\mu}(x), {}^{a}g_{5}{}^{a}\mathbf{A}_{DE}{}^{\mu}(x), {}^{a}g_{6}{}^{a}\mathbf{U}^{\mu}(x), {}^{a}g_{7}{}^{a}\mathbf{V}^{\mu}(x), {}^{a}g_{8}{}^{a}\mathbf{A}_{DSU(3)}{}^{\mu}(x_{C}), {}^{a}g_{9}{}^{a}\mathbf{U}_{\mathbf{D}}{}^{\mu}(x), {}^{a}g_{10}{}^{a}\mathbf{V}_{\mathbf{D}}{}^{\mu}(x)) \quad (4.1) $$

where a labels the layer, a = 1, 2, 3, 4. We label the respective coupling constants in each layer a as ${}^{a}g_{1}$, ${}^{a}g_{2}$, …, ${}^{a}g_{10}$. In the equation above: the subscript ‘D’ labels Dark matter interactions, ‘W’ labels Weak fields, ‘E’ labels Electromagnetic fields, U labels U(4) Generation group fields, ‘V’ labels U(4) Layer group fields, U_D labels Dark U(4) Generation group fields, and V_D labels Dark U(4) Layer group fields. A_S labels the U(4) Species Group fields.

---

[13] This section has equations obtained from Blaha (2017d) and (2018e).

We define the sum over a and the components of the vector ${}^{a}\mathbf{A}_{I}^{\mu}(x)$ labeled with i = 1, 2 for each of the paired PseudoQuantum fields, by

$$\mathbf{C}_{I}^{\mu}(x) = \Sigma_{a,i}\ {}^{a}\mathbf{A}_{Ii}^{\mu}(x) + \Sigma_{i}\ (g_{11}\mathbf{A}_{S}^{i\mu}(x) + g_{\Theta}\mathbf{A}_{\Theta}^{i\mu}(x)) \tag{4.2}$$

We begin by considering the case of one layer of vector bosons below omitting the ^{a} superscript. The generalization to four layers is straightforward.

Using the above definitions the *PseudoQuantum* covariant derivative of a 4-vector Z_μ is

$$\begin{aligned} D_\nu Z_\mu &= (\partial_\nu + iF_\nu)Z_\mu - H^{\sigma}{}_{\nu\mu}Z_\sigma \\ &= [g^{\sigma}{}_{\mu}\partial_\nu + ig^{\sigma}{}_{\mu}F_\nu - H^{\sigma}{}_{\nu\mu}]Z_\sigma \\ &= [g^{\sigma}{}_{\mu}\partial_\nu + iD^{\sigma}{}_{\mu\nu}]Z_\sigma \end{aligned} \tag{4.3}$$

where[14]

$$F^{\mu} = C_{I}^{1\mu}(x) + \mathbf{C}_{I}^{2\mu}(x) + A_{R}^{1\mu} + A_{R}^{2\mu} + B^{1\mu} + B^{2\mu} \tag{4.4}$$

and

$$\begin{aligned} H^{\sigma}{}_{\nu\mu} &= \Gamma_{GR}{}^{\sigma}{}_{\nu\mu} + \Gamma_{GR}{}^{2\sigma}{}_{\nu\mu} \\ D^{\sigma}{}_{\mu\nu} &= g^{\sigma}{}_{\mu}F_{\nu} + iH^{\sigma}{}_{\nu\mu} \end{aligned} \tag{4.5}$$

where we have abstracted the complex part of the complex affine connection into the U(4) gauge field A_S^{μ}. $H^{\sigma}{}_{\nu\mu}$ is the real-valued part of the complex affine connection.

We define the full vector gauge field covariant derivative to be

$$D^{\mu} = \partial^{\mu} - i(C_{I}^{1\mu}(x) + \mathbf{C}_{I}^{2\mu}(x)) \times \tag{4.5a}$$

Commutation relations of the vector fields in F_μ are implicit when the covariant derivative is applied to vectors and tensors such as Z_σ. This is indicated by the × symbol above.

---

[14] We will omit the insertion of the spinor coupling constants of the spinor connection $B^{1\mu}$ and $B^{2\mu}$ in eq. 11.2 in the interests of simplifying expressions.

4.2 The Curvature Tensor

The curvature tensor applied to a 4-vector Z_β is[15]

$$R'^{\beta}{}_{\sigma\nu\mu}Z_\beta = g^{\alpha}{}_{\mu}(\partial_\nu + iF_\nu)g^{\beta}{}_{\sigma}(\partial_\alpha + iF_\alpha)Z_\beta - H^{\alpha}{}_{\mu\nu}g^{\beta}{}_{\sigma}(\partial_\alpha + iF_\alpha)Z_\beta + \quad (4.6)$$
$$+ H^{\alpha}{}_{\mu\nu}H^{\beta}{}_{\sigma\alpha}Z_\beta - g^{\alpha}{}_{\mu}(\partial_\nu + iF_\nu)H^{\beta}{}_{\sigma\alpha}Z_\beta - H^{\gamma}{}_{\nu\sigma}\{g^{\alpha}{}_{\gamma}(\partial_\mu + iF_\mu)Z_\alpha - H^{\alpha}{}_{\gamma\mu}Z_\alpha\} -$$
$$- \{\mu\leftrightarrow\nu\}$$

$$= ig^{\beta}{}_{\sigma}(\partial_\nu F_\mu - \partial_\mu F_\nu - i[F_\nu, F_\mu])Z_\beta + (\partial_\mu H^{\beta}{}_{\sigma\nu} - \partial_\nu H^{\beta}{}_{\sigma\mu} + H^{\gamma}{}_{\nu\sigma}H^{\beta}{}_{\gamma\mu} - H^{\gamma}{}_{\mu\sigma}H^{\beta}{}_{\gamma\nu})Z_\beta$$

$$= ig^{\beta}{}_{\sigma}(F_{E}{}^{1}{}_{\nu\mu} + F_{E}{}^{2}{}_{\nu\mu} + F_{W}{}^{1}{}_{\nu\mu} + F_{W}{}^{2}{}_{\nu\mu} + F_{DE}{}^{1}{}_{\nu\mu} + F_{DE}{}^{2}{}_{\nu\mu} + F_{DW}{}^{1}{}_{\nu\mu} + F_{DW}{}^{2}{}_{\nu\mu} + F_{SU(3)}{}^{1}{}_{\nu\mu} + F_{SU(3)}{}^{2}{}_{\nu\mu} +$$
$$+ F_{DSU(3)}{}^{1}{}_{\nu\mu} + F_{DSU(3)}{}^{2}{}_{\nu\mu} + F_{U}{}^{1}{}_{\nu\mu} + F_{U}{}^{2}{}_{\nu\mu} + F_{V}{}^{1}{}_{\nu\mu} + F_{V}{}^{2}{}_{\nu\mu} + F_{UD}{}^{1}{}_{\nu\mu} + F_{UD}{}^{2}{}_{\nu\mu} + F_{VD}{}^{1}{}_{\nu\mu} + F_{VD}{}^{2}{}_{\nu\mu} +$$
$$+ F_{S}{}^{1}{}_{\nu\mu} + F_{S}{}^{2}{}_{\nu\mu} + F_{\Theta}{}^{1}{}_{\nu\mu} + F_{\Theta}{}^{2}{}_{\nu\mu} + F_{B}{}^{1}{}_{\nu\mu} + F_{B}{}^{2}{}_{\nu\mu} +$$
$$+ F_{A}{}^{1}{}_{\nu\mu} + F_{A}{}^{2}{}_{\nu\mu})Z_\beta + (\partial_\mu H^{\beta}{}_{\sigma\nu} - \partial_\nu H^{\beta}{}_{\sigma\mu} + H^{\gamma}{}_{\nu\sigma}H^{\beta}{}_{\gamma\mu} - H^{\gamma}{}_{\mu\sigma}H^{\beta}{}_{\gamma\nu})Z_\beta$$

$$= R'_{E}{}^{\beta}{}_{\sigma\nu\mu}Z_\beta + R'_{SU(2)}{}^{\beta}{}_{\sigma\nu\mu}Z_\beta + R'_{DE}{}^{\beta}{}_{\sigma\nu\mu}Z_\beta + R'_{DSU(2)}{}^{\beta}{}_{\sigma\nu\mu}Z_\beta + R'_{SU(3)}{}^{\beta}{}_{\sigma\nu\mu}Z_\beta +$$
$$+ R'_{DSU(3)}{}^{\beta}{}_{\sigma\nu\mu}Z_\beta + R'_{U}{}^{\beta}{}_{\sigma\nu\mu}Z_\beta + R'_{V}{}^{\beta}{}_{\sigma\nu\mu}Z_\beta + R'_{UD}{}^{\beta}{}_{\sigma\nu\mu}Z_\beta + R'_{VD}{}^{\beta}{}_{\sigma\nu\mu}Z_\beta + R'_{S}{}^{\beta}{}_{\sigma\nu}Z_\beta +$$
$$+ R'_{\Theta}{}^{\beta}{}_{\sigma\nu}Z_\beta + R'_{A}{}^{\beta}{}_{\sigma\nu}Z_\beta + R'_{B}{}^{\beta}{}_{\sigma\nu}Z_\beta + R'_{G}{}^{\beta}{}_{\sigma\nu\mu}Z_\beta$$

where all $F_{...}{}^{1}{}_{\nu\mu}$ and $F_{...}{}^{2}{}_{\nu\mu}$ terms have summations over the four layers (see below) except the terms[16] $F_{S}{}^{1}{}_{\nu\mu} + F_{S}{}^{2}{}_{\nu\mu} + F_{\Theta}{}^{1}{}_{\nu\mu} + F_{\Theta}{}^{2}{}_{\nu\mu} + F_{A}{}^{1}{}_{\nu\mu} + F_{A}{}^{2}{}_{\nu\mu} + F_{B}{}^{1}{}_{\nu\mu} + F_{B}{}^{2}{}_{\nu\mu}$, and where[17]

$$R'_{SU(3)}{}^{\beta}{}_{\sigma\nu\mu} = ig^{\beta}{}_{\sigma}(F_{SU(3)}{}^{1}{}_{\nu\mu} + F_{SU(3)}{}^{2}{}_{\nu\mu}) \quad (4.7)$$
$$R'_{SU(2)}{}^{\beta}{}_{\sigma\nu\mu} = ig^{\beta}{}_{\sigma}(F_{W}{}^{1}{}_{\nu\mu} + F_{W}{}^{2}{}_{\nu\mu})$$
$$R'_{E}{}^{\beta}{}_{\sigma\nu\mu} = ig^{\beta}{}_{\sigma}(F_{E}{}^{1}{}_{\nu\mu} + F_{E}{}^{2}{}_{\nu\mu})$$
$$R'_{U}{}^{\beta}{}_{\sigma\nu\mu} = ig^{\beta}{}_{\sigma}(F_{U}{}^{1}{}_{\nu\mu} + F_{U}{}^{2}{}_{\nu\mu})$$
$$R'_{V}{}^{\beta}{}_{\sigma\nu\mu} = ig^{\beta}{}_{\sigma}(F_{V}{}^{1}{}_{\nu\mu} + F_{V}{}^{2}{}_{\nu\mu})$$
$$R'_{DSU(3)}{}^{\beta}{}_{\sigma\nu\mu} = ig^{\beta}{}_{\sigma}(F_{DSU(3)}{}^{1}{}_{\nu\mu} + F_{DSU(3)}{}^{2}{}_{\nu\mu})$$

---

[15] **With an implicit summation over layers understood.**

[16] The U(4) General Relativity Reality Group fields $A_R{}^{\beta}$ have $F_{A}{}^{1}{}_{\nu\mu} + F_{A}{}^{2}{}_{\nu\mu}$.

[17] The B field is the General Relativistic spinor connection. Its effects are miniscule in physical situations except for extreme cases that have not as yet been encountered experimentally.

$$R'_{DSU(2)}{}^{\beta}{}_{\sigma\nu\mu} = ig^{\beta}{}_{\sigma}(F_{DW}{}^{1}{}_{\nu\mu} + F_{DW}{}^{2}{}_{\nu\mu})$$
$$R'_{DE}{}^{\beta}{}_{\sigma\nu\mu} = ig^{\beta}{}_{\sigma}(F_{DE}{}^{1}{}_{\nu\mu} + F_{DE}{}^{2}{}_{\nu\mu})$$
$$R'_{U_D}{}^{\beta}{}_{\sigma\nu\mu} = ig^{\beta}{}_{\sigma}(F_{U_D}{}^{1}{}_{\nu\mu} + F_{U_D}{}^{2}{}_{\nu\mu})$$
$$R'_{V_D}{}^{\beta}{}_{\sigma\nu\mu} = ig^{\beta}{}_{\sigma}(F_{V_D}{}^{1}{}_{\nu\mu} + F_{V_D}{}^{2}{}_{\nu\mu})$$
$$R'_{S}{}^{\beta}{}_{\sigma\nu\mu} = ig^{\beta}{}_{\sigma}(F_{S}{}^{1}{}_{\nu\mu} + F_{S}{}^{2}{}_{\nu\mu})$$
$$R'_{\Theta}{}^{\beta}{}_{\sigma\nu\mu} = ig^{\beta}{}_{\sigma}(F_{\Theta}{}^{1}{}_{\nu\mu} + F_{\Theta}{}^{2}{}_{\nu\mu})$$
$$R'_{B}{}^{\beta}{}_{\sigma\nu\mu} = ig^{\beta}{}_{\sigma}(F_{B}{}^{1}{}_{\nu\mu} + F_{B}{}^{2}{}_{\nu\mu})$$
$$R'_{A}{}^{\beta}{}_{\sigma\nu\mu} = ig^{\beta}{}_{\sigma}(F_{A}{}^{1}{}_{\nu\mu} + F_{A}{}^{2}{}_{\nu\mu})$$

and

$$\begin{aligned} R'_{G}{}^{\beta}{}_{\sigma\nu\mu} = {} & \partial_\mu H^{1\beta}{}_{\sigma\nu} - \partial_\nu H^{1\beta}{}_{\sigma\mu} + H^{1\gamma}{}_{\nu\sigma}H^{1\beta}{}_{\gamma\mu} - H^{1\gamma}{}_{\mu\sigma}H^{1\beta}{}_{\gamma\nu} + \partial_\mu H^{2\beta}{}_{\sigma\nu} - \partial_\nu H^{2\beta}{}_{\sigma\mu} + \\ & + H^{2\gamma}{}_{\nu\sigma}H^{2\beta}{}_{\gamma\mu} - H^{2\gamma}{}_{\mu\sigma}H^{2\beta}{}_{\gamma\nu} + H^{1\gamma}{}_{\nu\sigma}H^{2\beta}{}_{\gamma\mu} - H^{1\gamma}{}_{\mu\sigma}H^{2\beta}{}_{\gamma\nu} + H^{2\gamma}{}_{\nu\sigma}H^{1\beta}{}_{\gamma\mu} - \Gamma^{2\gamma}{}_{\mu\sigma}\Gamma^{\beta}{}_{\gamma\nu} \\ & = R^{1\beta}{}_{\sigma\nu\mu} + R^{2\beta}{}_{\sigma\nu\mu} \end{aligned} \tag{4.8}$$

with

$$\begin{aligned} H^{\beta}{}_{\sigma\nu\mu} &= \partial_\mu H^{\beta}{}_{\sigma\nu} - \partial_\nu H^{\beta}{}_{\sigma\mu} + H^{\gamma}{}_{\nu\sigma}H^{\beta}{}_{\gamma\mu} - H^{\gamma}{}_{\mu\sigma}H^{\beta}{}_{\gamma\nu} \\ R^{1\beta}{}_{\sigma\nu\mu} &= \partial_\mu H^{1\beta}{}_{\sigma\nu} - \partial_\nu H^{1\beta}{}_{\sigma\mu} + H^{1\gamma}{}_{\nu\sigma}H^{1\beta}{}_{\gamma\mu} - H^{1\gamma}{}_{\mu\sigma}H^{1\beta}{}_{\gamma\nu} \\ R^{2\beta}{}_{\sigma\nu\mu p} &= \partial_\mu H^{2\beta}{}_{\sigma\nu} - \partial_\nu H^{2\beta}{}_{\sigma\mu} + H^{2\gamma}{}_{\nu\sigma}H^{2\beta}{}_{\gamma\mu} - H^{2\gamma}{}_{\mu\sigma}H^{2\beta}{}_{\gamma\nu} + \\ & \quad + H^{1\gamma}{}_{\nu\sigma}H^{2\beta}{}_{\gamma\mu} - H^{1\gamma}{}_{\mu\sigma}H^{2\beta}{}_{\gamma\nu} + H^{2\gamma}{}_{\nu\sigma}H^{1\beta}{}_{\gamma\mu} - H^{2\gamma}{}_{\mu\sigma}H^{1\beta}{}_{\gamma\nu} \end{aligned} \tag{4.9}$$

and

$$\begin{aligned} H^{1\sigma}{}_{\nu\mu} &= \Gamma_{GR}{}^{\sigma}{}_{\nu\mu} \\ H^{2\sigma}{}_{\nu\mu} &= \Gamma_{GR}{}^{2\sigma}{}_{\nu\mu} \end{aligned} \tag{4.10}$$

and with summations over four layers indicated by Σ (Layer numbers on fields are not shown to avoid clutter.) As a result we have

$$\begin{aligned} F_{SU(3)}{}^{1}{}_{\varkappa\mu} &= \Sigma\ \{\partial A_{SU(3)}{}^{1}{}_{\mu}/\partial x^{\varkappa} - \partial A_{SU(3)}{}^{1}{}_{\varkappa}/\partial x^{\mu} + ig_1[A_{SU(3)}{}^{1}{}_{\varkappa}, A_{SU(3)}{}^{1}{}_{\mu}]\ \} \\ F_{W}{}^{1}{}_{\varkappa\mu} &= \Sigma\ \{\partial W^{1}{}_{\mu}/\partial x^{\varkappa} - \partial W^{1}{}_{\varkappa}/\partial x^{\mu} + ig_2[W^{1}{}_{\varkappa}, W^{1}{}_{\mu}]\ \} \\ F_{E}{}^{1}{}_{\varkappa\mu} &= \Sigma\ \{\partial A_{E}{}^{1}{}_{\mu}/\partial x^{\varkappa} - \partial A_{E}{}^{1}{}_{\varkappa}/\partial x^{\mu}\ \} \\ F_{DW}{}^{1}{}_{\varkappa\mu} &= \Sigma\ \{\partial W_{D}{}^{1}{}_{\mu}/\partial x^{\varkappa} - \partial W_{D}{}^{1}{}_{\varkappa}/\partial x^{\mu} + ig_4[W_{D}{}^{1}{}_{\varkappa}, W_{D}{}^{1}{}_{\mu}]\ \} \end{aligned} \tag{4.11}$$

$$F_{DE}{}^{1}{}_{\varkappa\mu} = \Sigma\ \{\partial A_{DE}{}^{1}{}_{\mu}/\partial x^{\varkappa} - \partial A_{DE}{}^{1}{}_{\varkappa}/\partial x^{\mu}\}$$
$$F_{DSU(3)}{}^{1}{}_{\varkappa\mu} = \Sigma\ \{\partial A_{DSU(3)}{}^{1}{}_{\mu}/\partial x^{\varkappa} - \partial A_{DSU(3)}{}^{1}{}_{\varkappa}/\partial x^{\mu} + ig_8[A_{DSU(3)}{}^{1}{}_{\varkappa}, A_{DSU(3)}{}^{1}{}_{\mu}]\ \}$$
$$F_{U}{}^{1}{}_{\varkappa\mu} = \Sigma\ \{\partial U^{1}{}_{\mu}/\partial x^{\varkappa} - \partial U^{1}{}_{\varkappa}/\partial x^{\mu} + ig_6[U^{1}{}_{\varkappa}, U^{1}{}_{\mu}]\ \}$$
$$F_{V}{}^{1}{}_{\varkappa\mu} = \Sigma\ \{\partial V^{1}{}_{\mu}/\partial x^{\varkappa} - \partial V^{1}{}_{\varkappa}/\partial x^{\mu} + ig_7[V^{1}{}_{\varkappa}, V^{1}{}_{\mu}]\ \}$$
$$F_{U_D}{}^{1}{}_{\kappa\mu} \equiv F_{DU}{}^{1}{}_{\varkappa\mu} = \Sigma\ \{\partial U_D{}^{1}{}_{\mu}/\partial x^{\kappa} - \partial U_D{}^{1}{}_{\kappa}/\partial x^{\mu} + ig_9[U_D{}^{1}{}_{\kappa}, U_D{}^{1}{}_{\mu}]\ \}$$
$$F_{V_D}{}^{1}{}_{\varkappa\mu} \equiv F_{DV}{}^{1}{}_{\varkappa\mu} = \Sigma\ \{\partial V_D{}^{1}{}_{\mu}/\partial x^{\varkappa} - \partial V_D{}^{1}{}_{\varkappa}/\partial x^{\mu} + ig_{10}[V_D{}^{1}{}_{\varkappa}, V_D{}^{1}{}_{\mu}]\ \}$$
$$F_{S}{}^{1}{}_{\varkappa\mu} = \partial A_S{}^{1}{}_{\mu}/\partial x^{\varkappa} - \partial A_S{}^{1}{}_{\varkappa}/\partial x^{\mu} + ig_{11}[A_S{}^{1}{}_{\varkappa}, A_S{}^{1}{}_{\mu}]$$
$$F_{\Theta}{}^{1}{}_{\varkappa\mu} = \partial A_{\Theta}{}^{1}{}_{\mu}/\partial x^{\varkappa} - \partial A_{\Theta}{}^{1}{}_{\varkappa}/\partial x^{\mu} + ig_{\Theta}[A_{\Theta}{}^{1}{}_{\varkappa}, A_{\Theta}{}^{1}{}_{\mu}]$$
$$F_{B}{}^{1}{}_{\varkappa\mu} = \partial B^{1}{}_{\mu}/\partial x^{\varkappa} - \partial B^{1}{}_{\varkappa}/\partial x^{\mu} + i[B^{1}{}_{\varkappa}, B^{1}{}_{\mu}]$$
$$F_{A}{}^{1}{}_{\varkappa\mu} = \partial A_R{}^{1}{}_{\mu}/\partial x^{\varkappa} - \partial A_R{}^{1}{}_{\varkappa}/\partial x^{\mu} + i[A_R{}^{1}{}_{\varkappa}, A_R{}^{1}{}_{\mu}]$$

$$F_{SU(3)}{}^{2}{}_{\varkappa\mu} = \Sigma\ \{\partial A_{SU(3)}{}^{2}{}_{\mu}/\partial x^{\varkappa} - \partial A_{SU(3)}{}^{2}{}_{\varkappa}/\partial x^{\mu} + ig_1[A_{SU(3)}{}^{2}{}_{\varkappa}, A_{SU(3)}{}^{2}{}_{\mu}] + ig_1[A_{SU(3)}{}^{1}{}_{\varkappa}, A_{SU(3)}{}^{2}{}_{\mu}] + ig_1[A_{SU(3)}{}^{2}{}_{\varkappa}, A_{SU(3)}{}^{1}{}_{\mu}]\} \qquad (4.12)$$

$$F_{W}{}^{2}{}_{\varkappa\mu} = \Sigma\ \{\partial W^{2}{}_{\mu}/\partial x^{\varkappa} - \partial W^{2}{}_{\varkappa}/\partial x^{\mu} + ig_2[W^{2}{}_{\varkappa}, W^{2}{}_{\mu}] + ig_2[W^{1}{}_{\varkappa}, W^{2}{}_{\mu}] + ig_2[W^{2}{}_{\varkappa}, W^{1}{}_{\mu}]\ \}$$
$$F_{E}{}^{2}{}_{\varkappa\mu} = \Sigma\ \{\partial A_E{}^{2}{}_{\mu}/\partial x^{\varkappa} - \partial A_E{}^{2}{}_{\varkappa}/\partial x^{\mu}\}$$

$$F_{DSU(3)}{}^{2}{}_{\varkappa\mu} = \Sigma\ \{\partial A_{DSU(3)}{}^{2}{}_{\mu}/\partial x^{\varkappa} - \partial A_{DSU(3)}{}^{2}{}_{\varkappa}/\partial x^{\mu} + ig_8[A_{DSU(3)}{}^{2}{}_{\varkappa}, A_{DSU(3)}{}^{2}{}_{\mu}] + ig_8[A_{DSU(3)}{}^{1}{}_{\varkappa}, A_{DSU(3)}{}^{2}{}_{\mu}] + ig_8[A_{DSU(3)}{}^{2}{}_{\varkappa}, A_{DSU(3)}{}^{1}{}_{\mu}]\}$$

$$F_{DW}{}^{2}{}_{\varkappa\mu} = \Sigma\ \{\partial W_D{}^{2}{}_{\mu}/\partial x^{\varkappa} - \partial W_D{}^{2}{}_{\varkappa}/\partial x^{\mu} + ig_4[W_D{}^{2}{}_{\varkappa}, W_D{}^{2}{}_{\mu}] + ig_4[W_D{}^{1}{}_{\varkappa}, W_D{}^{2}{}_{\mu}] + ig_4[W_D{}^{2}{}_{\varkappa}, W_D{}^{1}{}_{\mu}]\}$$

$$F_{DE}{}^{2}{}_{\varkappa\mu} = \Sigma\ \{\partial A_{DE}{}^{2}{}_{\mu}/\partial x^{\varkappa} - \partial A_{DE}{}^{2}{}_{\varkappa}/\partial x^{\mu}\}$$

$$F_{U}{}^{2}{}_{\varkappa\mu} = \Sigma\ \{\partial U^{2}{}_{\mu}/\partial x^{\varkappa} - \partial U^{2}{}_{\varkappa}/\partial x^{\mu} + ig_6[U^{2}{}_{\varkappa}, U^{2}{}_{\mu}] + ig_6[U^{1}{}_{\varkappa}, U^{2}{}_{\mu}] + ig_6[U^{2}{}_{\varkappa}, U^{1}{}_{\mu}]\}$$

$$F_{V\,\varkappa\mu}^{\;2} = \Sigma\ \{\partial V^2_{\ \mu}/\partial x^\varkappa - \partial V^2_{\ \varkappa}/\partial x^\mu + ig_7[V^2_{\ \varkappa}, V^2_{\ \mu}] + ig_7[V^1_{\ \varkappa}, V^2_{\ \mu}] + ig_7[V^2_{\ \varkappa}, V^1_{\ \mu}]\}$$

$$F_{U_D\,\varkappa\mu}^{\;2} = \Sigma\ \{\partial U_{D\ \mu}^{\ 2}/\partial x^\varkappa - \partial U_{D\ \varkappa}^{\ 2}/\partial x^\mu + ig_9[U_{D\ \varkappa}^{\ 2}, U_{D\ \mu}^{\ 2}] + ig_9[U_{D\ \varkappa}^{\ 1}, U_{D\ \mu}^{\ 2}] + + ig_6[U_{D\ \varkappa}^{\ 2}, U_{D\ \mu}^{\ 1}]\} \equiv F_{DU\ \varkappa\mu}^{\ \ 2}$$

$$F_{V_D\,\varkappa\mu}^{\;2} = \Sigma\ \{\partial V_{D\ \mu}^{\ 2}/\partial x^\varkappa - \partial V_{D\ \varkappa}^{\ 2}/\partial x^\mu + ig_7[V_{D\ \varkappa}^{\ 2}, V_{D\ \mu}^{\ 2}] + ig_7[V_{D\ \varkappa}^{\ 1}, V_{D\ \mu}^{\ 2}] + + ig_7[V_{D\ \varkappa}^{\ 2}, V_{D\ \mu}^{\ 1}]\} \equiv F_{DV\ \varkappa\mu}^{\ \ 2}$$

$$F_{S\ \kappa\mu}^{\ 2} = \partial A_{S\ \mu}^{\ 2}/\partial x^\kappa - \partial A_{S\ \kappa}^{\ 2}/\partial x^\mu + ig_8[A_{S\ \kappa}^{\ 2}, A_{S\ \mu}^{\ 2}] + ig_{11}[A_{S\ \kappa}^{\ 1}, A_{S\ \mu}^{\ 2}] + ig_{11}[A_{S\ \kappa}^{\ 2}, A_{S\ \mu}^{\ 1}]$$

$$F_{\Theta\ \kappa\mu}^{\ 2} = \partial A_{\Theta\ \mu}^{\ 2}/\partial x^\kappa - \partial A_{\Theta\ \kappa}^{\ 2}/\partial x^\mu + ig_\Theta[A_{\Theta\ \kappa}^{\ 2}, A_{\Theta\ \mu}^{\ 2}] + ig_\Theta[A_{\Theta\ \kappa}^{\ 1}, A_{\Theta\ \mu}^{\ 2}] + ig_\Theta[A_{\Theta\ \kappa}^{\ 2}, A_{\Theta\ \mu}^{\ 1}]$$

$$F_{B\ \varkappa\mu}^{\ 2} = \partial B^2_{\ \mu}/\partial x^\varkappa - \partial B^2_{\ \varkappa}/\partial x^\mu + i[B^2_{\ \mu}, B^2_{\ \varkappa}] + i[B^1_{\ \mu}, B^2_{\ \varkappa}] + i[B^2_{\ \mu}, B^1_{\ \varkappa}]$$

$$F_{A\ \varkappa\mu}^{\ 2} = \partial\ A_{R\ \mu}^{\ 2}/\partial x^\varkappa - \partial\ A_{R\ \varkappa}^{\ 2}/\partial x^\mu + i[A_{R\ \mu}^{\ 2}, A_{R\ \varkappa}^{\ 2}] + i[A_{R\ \mu}^{\ 1}, A_{R\ \varkappa}^{\ 2}] + i[A_{R\ \mu}^{\ 2}, A_{R\ \varkappa}^{\ 1}]$$

Note that $R'^{\beta}_{\ \sigma\nu\mu}$ factorizes into a

$$[U(1)\otimes SU(2)\otimes U(1)\otimes SU(2)\otimes SU(3)\ \otimes SU(3)\otimes U(4)\otimes U(4)]^4\otimes U(4)\otimes U(192)$$

part and a Riemann-Christoffel Gravitational curvature tensor part. For later use in defining a lagrangian we define

$$\begin{aligned} R'^{\beta}_{\ \sigma\nu\mu} = {} & R'_E{}^{1\beta}{}_{\sigma\nu\mu} + R'_E{}^{2\beta}{}_{\sigma\nu\mu} + R'_{SU(2)}{}^{1\beta}{}_{\sigma\nu\mu} + R'_{SU(2)}{}^{2\beta}{}_{\sigma\nu\mu} + R'_{DE}{}^{1\beta}{}_{\sigma\nu\mu} + R'_{DE}{}^{2\beta}{}_{\sigma\nu\mu} + \\ & + R'_{DSU(2)}{}^{1\beta}{}_{\sigma\nu\mu} + R'_{DSU(2)}{}^{2\beta}{}_{\sigma\nu\mu} + R'_{SU(3)}{}^{1\beta}{}_{\sigma\nu\mu} + R'_{SU(3)}{}^{2\beta}{}_{\sigma\nu\mu} + R'_{DSU(3)}{}^{1\beta}{}_{\sigma\nu\mu} + \\ & + R'_{DSU(3)}{}^{2\beta}{}_{\sigma\nu\mu} + R'_U{}^{1\beta}{}_{\sigma\nu\mu} + R'_U{}^{2\beta}{}_{\sigma\nu\mu} + R'_V{}^{1\beta}{}_{\sigma\nu\mu} + R'_V{}^{2\beta}{}_{\sigma\nu\mu} + R'_{DU}{}^{1\beta}{}_{\sigma\nu\mu} + \\ & + R'_{DU}{}^{2\beta}{}_{\sigma\nu\mu} + R'_{DV}{}^{1\beta}{}_{\sigma\nu\mu} + R'_{DV}{}^{2\beta}{}_{\sigma\nu\mu} + R'_S{}^{1\beta}{}_{\sigma\nu\mu} + R'_S{}^{2\beta}{}_{\sigma\nu\mu} + R'_\Theta{}^{1\beta}{}_{\sigma\nu\mu} + \\ & + R'_\Theta{}^{2\beta}{}_{\sigma\nu\mu} + R'_B{}^{1\beta}{}_{\sigma\nu\mu} + R'_B{}^{2\beta}{}_{\sigma\nu\mu} + R'_A{}^{1\beta}{}_{\sigma\nu\mu} + R'_A{}^{2\beta}{}_{\sigma\nu\mu} + R^{1\beta}{}_{\sigma\nu\mu} + R^{2\beta}{}_{\sigma\nu\mu} \end{aligned} \tag{4.13}$$

where

$$R'_{E}{}^{1\beta}{}_{\sigma\nu\mu} = ig^{\beta}{}_{\sigma}F_{E}{}^{1}{}_{\nu\mu} \qquad (4.14)$$
$$R'_{E}{}^{2\beta}{}_{\sigma\nu\mu} = ig^{\beta}{}_{\sigma}F_{DE}{}^{2}{}_{\nu\mu}$$
$$R'_{DE}{}^{1\beta}{}_{\sigma\nu\mu} = ig^{\beta}{}_{\sigma}F_{E}{}^{1}{}_{\nu\mu}$$
$$R'_{DE}{}^{2\beta}{}_{\sigma\nu\mu} = ig^{\beta}{}_{\sigma}F_{DE}{}^{2}{}_{\nu\mu}$$

$$R'_{SU(2)}{}^{1\beta}{}_{\sigma\nu\mu} = ig^{\beta}{}_{\sigma}F_{W}{}^{1}{}_{\nu\mu}$$
$$R'_{SU(2)}{}^{2\beta}{}_{\sigma\nu\mu} = ig^{\beta}{}_{\sigma}F_{DW}{}^{2}{}_{\nu\mu}$$

$$R'_{DSU(2)}{}^{1\beta}{}_{\sigma\nu\mu} = ig^{\beta}{}_{\sigma}F_{W}{}^{1}{}_{\nu\mu}$$
$$R'_{DSU(2)}{}^{2\beta}{}_{\sigma\nu\mu} = ig^{\beta}{}_{\sigma}F_{DW}{}^{2}{}_{\nu\mu}$$

$$R'_{SU(3)}{}^{1\beta}{}_{\sigma\nu\mu} = ig^{\beta}{}_{\sigma}F_{SU(3)}{}^{1}{}_{\nu\mu}$$
$$R'_{SU(3)}{}^{2\beta}{}_{\sigma\nu\mu} = ig^{\beta}{}_{\sigma}F_{SU(3)}{}^{2}{}_{\nu\mu}$$

$$R'_{DSU(3)}{}^{1\beta}{}_{\sigma\nu\mu} = ig^{\beta}{}_{\sigma}F_{DSU(3)}{}^{1}{}_{\nu\mu}$$
$$R'_{DSU(3)}{}^{2\beta}{}_{\sigma\nu\mu} = ig^{\beta}{}_{\sigma}F_{DSU(3)}{}^{2}{}_{\nu\mu}$$

$$R'_{U}{}^{1\beta}{}_{\sigma\nu\mu} = ig^{\beta}{}_{\sigma}F_{U}{}^{1}{}_{\nu\mu}$$
$$R'_{U}{}^{2\beta}{}_{\sigma\nu\mu} = ig^{\beta}{}_{\sigma}F_{U}{}^{2}{}_{\nu\mu}$$

$$R'_{V}{}^{1\beta}{}_{\sigma\nu\mu} = ig^{\beta}{}_{\sigma}F_{V}{}^{1}{}_{\nu\mu}$$
$$R'_{V}{}^{2\beta}{}_{\sigma\nu\mu} = ig^{\beta}{}_{\sigma}F_{V}{}^{2}{}_{\nu\mu}$$

$$R'_{DU}{}^{1\beta}{}_{\sigma\nu\mu} = ig^{\beta}{}_{\sigma}F_{DU}{}^{1}{}_{\nu\mu}$$
$$R'_{DU}{}^{2\beta}{}_{\sigma\nu\mu} = ig^{\beta}{}_{\sigma}F_{DU}{}^{2}{}_{\nu\mu}$$

$$R'_{DV}{}^{1\beta}{}_{\sigma\nu\mu} = ig^{\beta}{}_{\sigma}F_{DV}{}^{1}{}_{\nu\mu}$$
$$R'_{DV}{}^{2\beta}{}_{\sigma\nu\mu} = ig^{\beta}{}_{\sigma}F_{DV}{}^{2}{}_{\nu\mu}$$

$$R'_S{}^{1\beta}{}_{\sigma\nu\mu} = ig^\beta{}_\sigma F_S{}^1{}_{\nu\mu}$$
$$R'_S{}^{2\beta}{}_{\sigma\nu\mu} = ig^\beta{}_\sigma F_S{}^2{}_{\nu\mu}$$
$$R'_\Theta{}^{1\beta}{}_{\sigma\nu\mu} = ig^\beta{}_\sigma F_\Theta{}^1{}_{\nu\mu}$$
$$R'_\Theta{}^{2\beta}{}_{\sigma\nu\mu} = ig^\beta{}_\sigma F_\Theta{}^2{}_{\nu\mu}$$

$$R'_B{}^{1\beta}{}_{\sigma\nu\mu} = ig^\beta{}_\sigma B^1{}_{\nu\mu}$$
$$R'_B{}^{2\beta}{}_{\sigma\nu\mu} = ig^\beta{}_\sigma B^2{}_{\nu\mu}$$
$$R'_A{}^{1\beta}{}_{\sigma\nu\mu} = ig^\beta{}_\sigma A_R{}^1{}_{\nu\mu}$$
$$R'_A{}^{2\beta}{}_{\sigma\nu\mu} = ig^\beta{}_\sigma A_R{}^2{}_{\nu\mu}$$

The total Ricci tensor is

$$R'_{\sigma\mu} = R'^\beta{}_{\sigma\beta\mu} \tag{4.15}$$

$$\begin{aligned}= & iF_E{}^1{}_{\sigma\mu} + iF_E{}^2{}_{\sigma\mu} + iF_W{}^1{}_{\sigma\mu} + iF_W{}^2{}_{\sigma\mu} + iF_{DE}{}^1{}_{\sigma\mu} + iF_{DE}{}^2{}_{\sigma\mu} + iF_{DW}{}^1{}_{\sigma\mu} + iF_{DW}{}^2{}_{\sigma\mu} + iF_{SU(3)}{}^1{}_{\sigma\mu} + iF_{SU(3)}{}^2{}_{\sigma\mu} + \\
& + iF_{DSU(3)}{}^1{}_{\sigma\mu} + iF_{DSU(3)}{}^2{}_{\sigma\mu} + iF_U{}^1{}_{\sigma\mu} + iF_U{}^2{}_{\sigma\mu} + iF_V{}^1{}_{\sigma\mu} + iF_V{}^2{}_{\sigma\mu} + iF_{DU}{}^1{}_{\sigma\mu} + iF_{DU}{}^2{}_{\sigma\mu} + iF_{DV}{}^1{}_{\sigma\mu} + \\
& + iF_{DV}{}^2{}_{\sigma\mu} + iF_S{}^1{}_{\sigma\mu} + iF_S{}^2{}_{\sigma\mu} + iF_\Theta{}^1{}_{\sigma\mu} + iF_\Theta{}^2{}_{\sigma\mu} + iF_B{}^1{}_{\sigma\mu} + iF_B{}^2{}_{\sigma\mu} + \\
& + iF_A{}^1{}_{\sigma\mu} + iF_A{}^2{}_{\sigma\mu} + \partial_\mu H^{1\beta}{}_{\sigma\beta} - \partial_\beta H^{1\beta}{}_{\sigma\mu} + H^{1\gamma}{}_{\beta\sigma} H^{1\beta}{}_{\gamma\mu} - H^{1\gamma}{}_{\mu\sigma} H^{1\beta}{}_{\gamma\beta} + \\
& + \partial_\mu H^{2\beta}{}_{\sigma\beta} - \partial_\beta H^{2\beta}{}_{\sigma\mu} + H^{2\gamma}{}_{\beta\sigma} H^{2\beta}{}_{\gamma\mu} - H^{2\gamma}{}_{\mu\sigma} H^{2\beta}{}_{\gamma\beta} + H^{1\gamma}{}_{\beta\sigma} H^{2\beta}{}_{\gamma\mu} - H^{1\gamma}{}_{\mu\sigma} H^{2\beta}{}_{\gamma\beta} + \\
& + H^{2\gamma}{}_{\beta\sigma} H^{1\beta}{}_{\gamma\mu} - H^{2\gamma}{}_{\mu\sigma} H^{1\beta}{}_{\gamma\beta}\end{aligned}$$

$$\begin{aligned}= & R'_E{}^1{}_{\sigma\mu} + R'_E{}^2{}_{\sigma\mu} + R'_{SU(2)}{}^1{}_{\sigma\mu} + R'_{SU(2)}{}^2{}_{\sigma\mu} + R'_{DE}{}^1{}_{\sigma\mu} + R'_{DE}{}^2{}_{\sigma\mu} + R'_{DSU(2)}{}^1{}_{\sigma\mu} + R'_{DSU(2)}{}^2{}_{\sigma\mu} + \\
& + R'_{SU(3)}{}^1{}_{\sigma\mu} + R'_{SU(3)}{}^2{}_{\sigma\mu} + R'_U{}^1{}_{\sigma\mu} + R'_U{}^2{}_{\sigma\mu} + R'_V{}^1{}_{\sigma\mu} + R'_V{}^2{}_{\sigma\mu} + R'_{DSU(3)}{}^1{}_{\sigma\mu} + R'_{DSU(3)}{}^2{}_{\sigma\mu} + \\
& + R'_{DU}{}^1{}_{\sigma\mu} + R'_{DU}{}^2{}_{\sigma\mu} + R'_{DV}{}^1{}_{\sigma\mu} + R'_{DV}{}^2{}_{\sigma\mu} + R'_S{}^1{}_{\sigma\mu} + R'_S{}^2{}_{\sigma\mu} + R'_\Theta{}^1{}_{\sigma\mu} + \\
& + R'_\Theta{}^2{}_{\sigma\mu} + R'_A{}^{1\beta}{}_{\sigma\beta\mu} + R'_A{}^{2\beta}{}_{\sigma\beta\mu} + R'_B{}^{1\beta}{}_{\sigma\beta\mu} + R'_B{}^{2\beta}{}_{\sigma\beta\mu} + R^1{}_{\sigma\mu} + R^2{}_{\sigma\mu} \\
= & R'^1{}_{\sigma\mu} + R'^2{}_{\sigma\mu}\end{aligned}$$

where

$$\begin{aligned}R'^1{}_{\sigma\mu} = & R'_E{}^1{}_{\sigma\mu} + R'_{SU(2)}{}^1{}_{\sigma\mu} + R'_{DE}{}^1{}_{\sigma\mu} + R'_{DSU(2)}{}^1{}_{\sigma\mu} + R'_{SU(3)}{}^1{}_{\sigma\mu} + R'_U{}^1{}_{\sigma\mu} + R'_V{}^1{}_{\sigma\mu} + \\
& + R'_{DSU(3)}{}^1{}_{\sigma\mu} + R'_{DU}{}^1{}_{\sigma\mu} + R'_{DV}{}^1{}_{\sigma\mu} + R'_S{}^1{}_{\sigma\mu} + R'_\Theta{}^1{}_{\sigma\mu} + R'_A{}^{1\beta}{}_{\sigma\beta\mu} + R'_B{}^{1\beta}{}_{\sigma\beta\mu} + R^1{}_{\sigma\mu}\end{aligned} \tag{4.16}$$

$$R'^{2}{}_{\sigma\mu} = R'_E{}^2{}_{\sigma\mu} + R'_{SU(2)}{}^2{}_{\sigma\mu} + R'_{DE}{}^2{}_{\sigma\mu} + R'_{DSU(2)}{}^2{}_{\sigma\mu} + R'_{SU(3)}{}^2{}_{\sigma\mu} + R'_U{}^2{}_{\sigma\mu} + R'_V{}^2{}_{\sigma\mu} + \\ + R'_{DSU(3)}{}^2{}_{\sigma\mu} + R'_{DU}{}^2{}_{\sigma\mu} + R'_{DV}{}^2{}_{\sigma\mu} + R'_S{}^2{}_{\sigma\mu} + R'_\Theta{}^2{}_{\sigma\mu} + R'_A{}^{2\beta}{}_{\sigma\beta\mu} + R'_B{}^{2\beta}{}_{\sigma\beta\mu} + R^2{}_{\sigma\mu} \tag{4.17}$$

with

$$\begin{aligned}
R'_E{}^1{}_{\sigma\mu} &= iF_E{}^1{}_{\sigma\mu} \\
R'_E{}^2{}_{\sigma\mu} &= iF_E{}^2{}_{\sigma\mu} \\
R'_{SU(2)}{}^1{}_{\sigma\mu} &= iF_W{}^1{}_{\sigma\mu} \\
R'_{SU(2)}{}^2{}_{\sigma\mu} &= iF_W{}^2{}_{\sigma\mu} \\
R'_{DE}{}^1{}_{\sigma\mu} &= iF_{DE}{}^1{}_{\sigma\mu} \\
R'_{DE}{}^2{}_{\sigma\mu} &= iF_{DE}{}^2{}_{\sigma\mu} \\
R'_{DSU(2)}{}^1{}_{\sigma\mu} &= iF_{DW}{}^1{}_{\sigma\mu} \\
R'_{DSU(2)}{}^2{}_{\sigma\mu} &= iF_{DW}{}^2{}_{\sigma\mu} \\
R'_{SU(3)}{}^1{}_{\sigma\mu} &= iF_{SU(3)}{}^1{}_{\sigma\mu} \\
R'_{SU(3)}{}^2{}_{\sigma\mu} &= iF_{SU(3)}{}^2{}_{\sigma\mu} \\
R'_U{}^1{}_{\sigma\mu} &= iF_U{}^1{}_{\sigma\mu} \\
R'_U{}^2{}_{\sigma\mu} &= iF_U{}^2{}_{\sigma\mu} \\
R'_V{}^1{}_{\sigma\mu} &= iF_V{}^1{}_{\sigma\mu} \\
R'_V{}^2{}_{\sigma\mu} &= iF_V{}^2{}_{\sigma\mu} \\
R'_{DSU(3)}{}^1{}_{\sigma\mu} &= iF_{DSU(3)}{}^1{}_{\sigma\mu} \\
R'_{DSU(3)}{}^2{}_{\sigma\mu} &= iF_{DSU(3)}{}^2{}_{\sigma\mu} \\
R'_{DU}{}^1{}_{\sigma\mu} &= iF_{DU}{}^1{}_{\sigma\mu} \\
R'_{DU}{}^2{}_{\sigma\mu} &= iF_{DU}{}^2{}_{\sigma\mu} \\
R'_{DV}{}^1{}_{\sigma\mu} &= iF_{DV}{}^1{}_{\sigma\mu} \\
R'_{DV}{}^2{}_{\sigma\mu} &= iF_{DV}{}^2{}_{\sigma\mu} \\
R'_S{}^1{}_{\sigma\mu} &= iF_S{}^1{}_{\sigma\mu} \\
R'_S{}^2{}_{\sigma\mu} &= iF_S{}^2{}_{\sigma\mu} \\
R'_\Theta{}^1{}_{\sigma\mu} &= iF_\Theta{}^1{}_{\sigma\mu} \\
R'_\Theta{}^2{}_{\sigma\mu} &= iF_\Theta{}^2{}_{\sigma\mu} \\
R'_A{}^1{}_{\sigma\mu} &= iF_B{}^1{}_{\sigma\mu}
\end{aligned} \tag{4.18}$$

$$R'_A{}^2{}_{\sigma\mu} = iF_B{}^2{}_{\sigma\mu}$$
$$R'_B{}^1{}_{\sigma\mu} = iF_B{}^1{}_{\sigma\mu}$$
$$R'_B{}^2{}_{\sigma\mu} = iF_B{}^2{}_{\sigma\mu}$$

with the further definition of $R''^1{}_{\sigma\mu}$ and $R''^2{}_{\sigma\mu}$:

$$R''^1{}_{\sigma\mu} = R'_{SU(3)}{}^1{}_{\sigma\mu} + R'_{DSU(3)}{}^1{}_{\sigma\mu} + R^1{}_{\sigma\mu}$$
$$R''^2{}_{\sigma\mu} = R'_{SU(3)}{}^2{}_{\sigma\mu} + R'_{DSU(3)}{}^2{}_{\sigma\mu} + R^2{}_{\sigma\mu} \tag{4.19}$$

$R'^1{}_{\sigma\mu}$ is the Ricci tensor. An additional Ricci-like tensor is

$$H_{\sigma_\mu} = H^\beta{}_{\sigma\beta\mu} \tag{4.20}$$

The curvature scalar is

$$R' = g^{\sigma\mu}R'_{\sigma\mu} = + \partial^\sigma H^{1\beta}{}_{\sigma\beta} - \partial_\beta H^{1\beta}{}_\sigma{}^\sigma + H^{1\gamma}{}_{\beta\sigma}H^{1\beta}{}_\gamma{}^\sigma - H^{1\gamma}{}_{\mu\sigma}H^{1\beta}{}_{\gamma\beta} + \partial^\sigma H^{2\beta}{}_{\sigma\beta} - \partial_\beta H^{2\beta}{}_\sigma{}^\sigma +$$
$$+ H^{2\gamma}{}_{\beta\sigma}H^{2\beta}{}_\gamma{}^\sigma - H^{2\gamma\sigma}{}_\sigma H^{2\beta}{}_{\gamma\beta} + H^{1\gamma}{}_{\beta\sigma}H^{2\beta}{}_\gamma{}^\sigma - H^{1\gamma\sigma}{}_\sigma H^{2\beta}{}_{\gamma\beta} + H^{2\gamma}{}_{\beta\sigma}H^{1\beta}{}_\gamma{}^\sigma -$$
$$- H^{2\gamma\sigma}{}_\sigma H^{1\beta}{}_{\gamma\beta} \tag{4.21}$$

$$= g^{\sigma\mu}(R^{1\beta}{}_{\sigma\beta\mu} + R^{2\beta}{}_{\sigma\beta\mu})$$

4.3 Vector Boson and Graviton Lagrangian Terms

We choose the vector boson and gravitational part of the lagrangian[18] of the Unified SuperStandard Theory (with the Higgs sector and the Faddeev-Popov terms gauge sector not displayed here) to be:

$$\mathcal{L} = \text{Tr}\,\sqrt{g}[MD_\nu R''^1{}_{\sigma\mu}D^\nu R''^{2\sigma\mu} + aR'^1{}_{\sigma\mu}R'^{2\sigma\mu} + bR' + cg^{\sigma\mu}g^2{}_{\sigma\mu} + c'g^{2\sigma\mu}g^2{}_{\sigma\mu} -$$
$$- dA_{SU(3)}{}^2{}_\mu A_{SU(3)}{}^{2\mu}] \tag{4.22}$$

---

[18] The rationale for this choice is 1) to obtain the known Stanard Model interactions, 2) to obtain a canonical formulation for this higher derivative theory, and 3) to introduce higher derivative terms that yield quark confinement and a theory of gravity that accounts for known deviations from Newtonian gravity such as described by MoND. See Blaha (2019g) and (2018e) and earlier books for details on these points.

with a sum over layers understood, where D_ν is given by eq. 4.5a, where M, a, b, c, c', and d are constants, and $R''^{i}{}_{\sigma\mu}$ for i = 1, 2 determined above.[19]

This higher derivative lagrangian maintains the locality of the theory but does entail a modest modification in the derivation of the Euler-Lagrange equations of motion. It also requires the use of principal value propagators rather than ordinary Feynman propagators for gluon and graviton interactions. Thus the Strong Interaction sector, and the Gravitation sector are Action-at-a-Distance theories that are similar in spirit to Wheeler-Feynman Electrodynamics. The two U(1) Electromagnetic sectors, the Generation group U(4) gauge field sector, the Layer group U(4) gauge field sector, the two SU2) Weak sectors, the U(4) A_S gauge field sector, the spinor connection sector, and the Θ-interaction sector may, or may not, be Action-at-a-Distance fields. They are not constrained to be Action-at-a-Distance by the present considerations.

Since we wish to apply our theory cosmologically, and within hadrons, where the gravitational spinor connections are negligible due to the smallness of the

---

[19] One may ask why $R''^{1}{}_{\sigma\mu}$ and $R''^{2}{}_{\sigma\mu}$ appear in the first term of the lagrangian, and not other interaction terms. We believe the primary reason is: "The extended vierbein $l^{\mu ai}(x)$ can be viewed as located at a point in a higher dimensional complex-valued space.

$$l^{\mu ai}(x) = (\partial \xi_X{}^{ai}(x)/\partial x_\mu)_{X=h(x)}$$

where $\xi_X{}^{ai}$ is a set of locally inertial coordinates located at point X, and x = h(x) is a 4-dimensional point in a tangent subspace of the higher dimensional space:

$$X = h(x)$$

The relation between complex 4-dimensional coordinates x and the higher dimensional coordinates X is an embedding of a 4-dimensional surface within the higher dimensional complex space when account is taken of the range of possible x values. We have considered such embeddings in Blaha (2015a), and in earlier books, and developed a theory of a higher dimensional complex-valued space (the *Megaverse*) that contains our universe and probably many other universes." Thus SU(3) and Gravitation have a special role in our particle dynamics based on geometry. The second reason is the common feature of color SU(3) and real-valued General Relativity is that they are the only interactions that do not participate in 'rotations of interactions' as described earlier and in chapter 31 of Blaha (2017b). The third, practical reason is the experimental reality that the Strong Interaction and Gravitation are known to have 'anomalous' features that will be seen to be remedied by these insertions while the other interactions are 'conventional.'

gravitational constant G and the 'smallness' of Gravitational B spinor connection effects on the cosmological scale, we set $F^1{}_{\nu\mu} = F^2{}_{\nu\mu} = 0$ and find[20]

$$\mathcal{L} = \mathrm{Tr}\ \sqrt{g}[MD_\nu(R'^1{}_{SU(3)\sigma\mu} + R'^1{}_{DSU(3)\sigma\mu} + R^1{}_{\sigma\mu})D^\nu(R'_{SU(3)}{}^{2\sigma\mu} + R'_{DSU(3)}{}^{2\sigma\mu} + R^{2\sigma\mu}) + + aR'^1{}_{\sigma\mu}R'^{2\sigma\mu} + bR' + cg^{\sigma\mu}g^2{}_{\sigma\mu} + c'g^{2\sigma\mu}g^2{}_{\sigma\mu} - dA_{SU(3)}{}^2{}_\mu A_{SU(3)}{}^{2\mu}] \quad (4.23)$$

Since there are no strong interaction fields in 'empty' space and gravity is negligible within hadrons,[21] we can drop the interaction terms between the Strong interaction and the Gravity interaction. However, we cannot drop the interaction terms amongst Electromagnetism, the Weak interaction, the Strong Interaction, the Generation groups U(4) interactions, the U(4) Layer groups interactions, the U(4) Species group interaction, and the U(192) Θ-interaction – within, and between, hadrons. The interaction terms between Electromagnetism and Gravitation are important cosmologically.

The above lagrangian terms can therefore be expressed as:[22]

$$\mathcal{L} = \mathcal{L}_E + \mathcal{L}_{SU(2)} + \mathcal{L}_{DE} + \mathcal{L}_{DSU(2)} + \mathcal{L}_{SU(3)} + \mathcal{L}_{DSU(3)} + \mathcal{L}_U + \mathcal{L}_V + \mathcal{L}_{DU} + \mathcal{L}_{DV} + \mathcal{L}_S + + \mathcal{L}_\Theta + \mathcal{L}_G + \mathcal{L}_{int} \quad (4.24)$$

where taking traces of $\mathcal{L}$s terms is understood, and with coupling constants not displayed below to avoid clutter,

$$\mathcal{L}_E = \mathrm{Tr}\ \sqrt{g}\{M\{[\partial_\nu + i(A_E{}^1{}_\nu + A_E{}^2{}_\nu)]F^1{}_{E\sigma\mu}[\partial^\nu + i(A_E{}^{1\nu} + A_E{}^{2\nu})]F^2{}_E{}^{\sigma\mu}\} + aF_E{}^1{}_{\sigma\mu}F_E{}^{2\sigma\mu}\} \quad (4.25)$$

$$\mathcal{L}_{SU(2)} = \mathrm{Tr}\ \sqrt{g}[aF_W{}^1{}_{\sigma\mu}F_W{}^{2\sigma\mu}]$$

[20] The constants have the dimensions: M has the dimension of inverse mass squared, b has dimension mass squared, a is dimensionless, c and c' have dimension mass, and d has dimension mass squared.

[21] We show gravity weakens at very short distances using our Two-Tier Quantum Field Theory formalism. See Appendix A, and Blaha (2003) and (2005a) among other books by the author.

[22] We only consider the gauge field lagrangian terms.

$$\mathcal{L}_{DE} = \mathrm{Tr}\,\sqrt{g}\{M\{[\partial_\nu + i(A_{DE}{}^1{}_\nu + A_{DE}{}^2{}_\nu)]F^1{}_{DE\sigma\mu}[\partial^\nu + i(A_{DE}{}^{1\nu} + A_{DE}{}^{2\nu})]F_{DE}{}^{2\sigma\mu}\} + \\ + aF_{DE}{}^1{}_{\sigma\mu}F_{DE}{}^{2\sigma\mu}\}$$

$$\mathcal{L}_{DSU(2)} = \mathrm{Tr}\,\sqrt{g}[aF_W{}^1{}_{\sigma\mu}F_W{}^{2\sigma\mu}]$$

$$\mathcal{L}_{SU(3)} = \mathrm{Tr}\,\sqrt{g}\{M[\partial_\nu + i(A_{SU(3)}{}^1{}_\nu + A_{SU(3)}{}^2{}_\nu)]F_{SU(3)}{}^1{}_{\sigma\mu}[\partial^\nu + i(A_{SU(3)}{}^{1\nu} + \\ + A_{SU(3)}{}^{2\nu})]F_{SU(3)}{}^{2\sigma\mu} + aF_{SU(3)}{}^1{}_{\sigma\mu}F_{SU(3)}{}^{2\sigma\mu} - dA_{SU(3)}{}^2{}_\mu A_{SU(3)}{}^{2\mu}\}$$

$$\mathcal{L}_{DSU(3)} = \mathrm{Tr}\,\sqrt{g}\{M[\partial_\nu + i(A_{DSU(3)}{}^1{}_\nu + A_{DSU(3)}{}^2{}_\nu)]F_{DSU(3)}{}^1{}_{\sigma\mu}[\partial^\nu + i(A_{DSU(3)}{}^{1\nu} + \\ + A_{DSU(3)}{}^{2\nu})]F_{DSU(3)}{}^{2\sigma\mu} + aF_{DSU(3)}{}^1{}_{\sigma\mu}F_{DSU(3)}{}^{2\sigma\mu} - dA_{DSU(3)}{}^2{}_\mu A_{DSU(3)}{}^{2\mu}\}$$

$$\mathcal{L}_U = \mathrm{Tr}\,\sqrt{g}[aF_U{}^1{}_{\sigma\mu}F_U{}^{2\sigma\mu}]$$

$$\mathcal{L}_V = \mathrm{Tr}\,\sqrt{g}[aF_V{}^1{}_{\sigma\mu}F_V{}^{2\sigma\mu}]$$

$$\mathcal{L}_{DU} = \mathrm{Tr}\,\sqrt{g}[aF_{DU}{}^1{}_{\sigma\mu}F_{DU}{}^{2\sigma\mu}]$$

$$\mathcal{L}_{DV} = \mathrm{Tr}\,\sqrt{g}[aF_{DV}{}^1{}_{\sigma\mu}F_{DV}{}^{2\sigma\mu}]$$

$$\mathcal{L}_S = \mathrm{Tr}\,\sqrt{g}[aF_S{}^1{}_{\sigma\mu}F_S{}^{2\sigma\mu}]$$

$$\mathcal{L}_\Theta = \mathrm{Tr}\,\sqrt{g}[aF_\Theta{}^1{}_{\sigma\mu}F_\Theta{}^{2\sigma\mu}]$$

$$\mathcal{L}_G = \mathrm{Tr}\,\sqrt{g}[MD_\nu R^1{}_{\sigma\mu}D^\nu R^{2\sigma\mu} + aR^1{}_{\sigma\mu}R^{2\sigma\mu} + bg^{\sigma\mu}(R^{1\beta}{}_{\sigma\beta\mu} + R^{2\beta}{}_{\sigma\beta\mu}) + cg^{\sigma\mu}g^2{}_{\sigma\mu} + c'g^{2\sigma\mu}g^2{}_{\sigma\mu}] \\ = \mathrm{Tr}\,\sqrt{g}[MD_\nu R^1{}_{\sigma\mu}D^\nu R^{2\sigma\mu} + aR^1{}_{\sigma\mu}R^{2\sigma\mu} + bH + cg^{\sigma\mu}g^2{}_{\sigma\mu} + c'g^{2\sigma\mu}g^2{}_{\sigma\mu}]$$

$$\mathcal{L}_{int} = \mathcal{L} - (\mathcal{L}_E + \mathcal{L}_{SU(2)} + \mathcal{L}_{DE} + \mathcal{L}_{DSU(2)} + \mathcal{L}_{SU(3)} + \mathcal{L}_U + \mathcal{L}_V + \mathcal{L}_{DSU(3)} + \mathcal{L}_{DU} + \mathcal{L}_{DV} + \mathcal{L}_S + \\ + \mathcal{L}_\Theta + \mathcal{L}_G) \qquad (4.26)$$

with appropriate sums over layers and gravitational B spinor connection terms omitted. Thus $\mathcal{L}_{SU(3)}$, $\mathcal{L}_{SU(2)}$, $\mathcal{L}_E$, $\mathcal{L}_{DE}$, $\mathcal{L}_{DSU(2)}$, $\mathcal{L}_{U}$,, $\mathcal{L}_V$, $\mathcal{L}_S$, $\mathcal{L}_\Theta$, and parts of $\mathcal{L}_{int}$ are the dominant interactions within hadrons, and $\mathcal{L}_G$, $\mathcal{L}_E$ and parts of $\mathcal{L}_{int}$ are the dominant interactions in space within the framework of this discussion.

The $D_\nu R^1{}_{\sigma\mu}$ and $D^\nu R^{2\sigma\mu}$ terms have the form:

$$D_\nu R^i{}_{\sigma\mu} = + \partial_\nu R^i{}_{\sigma\mu} \; - H^{1\beta}{}_{\sigma\nu} R^i{}_{\beta\mu} - H^{1\beta}{}_{\nu\mu} R^i{}_{\sigma\beta} \qquad (4.27)$$

for i = 1, 2 while covariant derivatives for internal symmetries are given by eq. 4.5a.

Blaha (2019g) and (2018e) discusses further details of the lagrangian obtained from the Riemann-Christoffel tensor, and its gravitation and Strong Interaction terms. The Dark Strong interaction terms generated by the first term in eq. 4.23 has an effect that parallels the Strong interaction case and leads to the confinement of Dark quarks.[23]

---

[23] See Blaha (2019g) and (2018e).

Space, Time, Symmetry in Our Universe

5. Quaternion Unified SuperStandard Theory (QUeST)

Our work on the Unified SuperStandard Theory, and in previous books, has uncovered a close similarity between the internal symmetries of the Standard Model sector and subgroups of the Lorentz group. They both exhibit U(1), SU(2) and SU(3) symmetries. In addition we showed Lorentz group boosts have four varieties that map nicely to the four types of species of the fundamental fermions: charged lepton, neutral lepton, up-type quarks and down-type quarks.

The similarities and success in understanding the origin of the four species of fundamental fermions led us to consider the possibility that a wider space might encompass both space-time and the internal symmetry groups in a manner that does not violate known "NoGo" theorems.

We found the Unified SuperStandard Theory was properly based on the Complex Lorentz group (and thus Complex General Relativity). Thus it seemed reasonable to consider a more general platform. Quaternions are the next step after complex numbers.

Quaternions have significant properties that distinguish them:

1.They are associative.

2. They are one of the two finite dimensional division rings having the real numbers as a proper subring. (The other is octonions—considered in chapter 13.)

3. They are non-commutative. (This is not a roadblock for quantum field theory which is also non-commutative in general.)

These features support the development of physics theories.[24]

5.1 Some Basic Quaternion Features

A quaternion is a 4-tuple of real numbers:

$$x = a + bi + jc + kd \quad = a + \mathbf{v} \tag{5.1}$$

where a, b, c, d are real numbers, and **v** is a 3-vector. A quaternion norm is defined by

$$||\mathbf{x}|| = \text{sqrt}(a^2 + b^2 + c^2 + d^2) \tag{5.2}$$

and the norm of **v** is

$$||\mathbf{v}|| = \text{sqrt}(\ b^2 + c^2 + d^2\ ) \tag{5.3}$$

A critically important identity is

$$e^x = e^a\ (\cos\ (||\mathbf{v}||) + \mathbf{v}/||\mathbf{v}||\ \sin(||\mathbf{v}||))s \tag{5.4}$$

.It will be used to define boosts in quaternion space.

5.2 Motivation and Procedure

Our goal is to create a larger dimension space within which we can derive our space-time and the Unified Superstandard Theory in such a way as to understand the similarity of Lorentz subgroups and Standard Model internal symmetry groups. The development of a deeper basis for the Unified SuperStandard Theory will lead to refinements in the theory.

There are two possible procedures to follow in developing the deeper basis:

---

[24] There is an extensive literature on quaternions starting with the original work of Hamilton. Some recent, relevant papers are: S. L. Adler, "Generalized Quantum Dynamics", IASSNS –HEP-93/32 (1993); S. De Leo, arXiv:hep-th/9506179 (1995); Rolf Dahm, arXiv:hep-th/9601207 (1996); S. De Leo, arXiv:hep-th/9508011 (1995); S. L. Adler, arXiv:hep-th/9607008 (1996) and references therein.

1. One can develop the Quantum Mechanics and Quantum Field Theory in a quaternion space and then extract the dynamics, fermion spectrum, gauge fields, and so on of our familiar space-time.

2. One can define a quaternion space, and then using its coordinates, directly extract the space-time, internal symmetries, fermion spectrum, gauge field spectrum and dynamics.

We have chosen the latter approach as it will directly lead to the Unified SuperStandard Theory.

In developing the deeper space, upon which we will build, we will take guidance from the derivation of the Unified SuperStandard Theory. This theory assumes a complex 4-dimensional space-time upon which Complex General Relativity is constructed. It then proceeds to complex flat space-time and Complex Relativity.

After defining features of Complex Lorentz transformations the Unified SuperStandard Theory uses Lorentz boosts to derive the four Dirac forms of the four fermion species. The boosts are required to boost a fermion from a rest state to a state with a real-valued energy, and a real or complex-valued 3-momenta. Thus a real time–complex-valued spatial part is required for the proper definition of species.

The Unified SuperStandard Theory then showed Lorenz subgroups mapped to Standard Model internal symmetry subgroups.

5.3 Definition of Quaternion Space – Biquaternion

Following the above stated procedure we define a biquaternion space with one "time" biquarternion and three "spatial" biquaternions.believing the 3+1 space-time of our experience is a reflection of this deeper level.

Time Biquaternion

$$t = (a + bi + jc + kd) + I(a' + b'i' + j'c' + k'd')$$

Spatial Biquaternions

$$x = (a_x + b_x i + jc_x + kd_x) + I(a'_x + b_x'i' + j'c_x' + k'd_x')$$

$$y = (a_y + b_y i + jc_y + kd_y) + I(a'_y + b_y'i' + j'c_y' + k'd_y')$$
$$z = (a_z + b_z i + jc_z + kd_z) + I(a'_z + b_z'i' + j'c_z' + k'd_z')$$

where I is another "imaginary" number with $I^2 = -1$ that introduces a further complexification leading to biquaternions.

Fig 5.1 symbolically depicts the space with a circle for each real-valued coordinate.

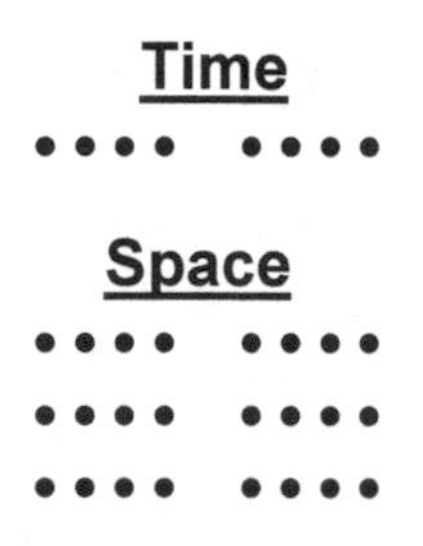

Figure 5.1. Four-Dimensional biquaternion space with coordinates represented by • 's.

The biquaternion space has 32 real dimensions (16 complex dimensions.).

5.4 Biquaternion Lorentz Group

Our definition of time and space biquaternion coordinates purposefully resembles those of our real space-time. One might ask why should there be a Lorentz-like group for biquaternion space? The only apparent reason is the need for a special speed c in our space-time. Without c the group of the above coordinates is the fundamental representation of the biquaternion U(4) group. In this group one cannot uniquely identify a particle rest frame. The biquaternion Lorentz transformations do have a unique speed c (the speed of light) and specify a unique rest frame for any particle. Thus we must select the Biquaternionic Lorentz group for biquaternion space to have physically required, unique particle rest frames.

Flat space biquaternion Special Relativity generalizes directly to a biquaternion General Relativity which may be constructed directly (mindful of quaternion non-commutativity).

The flat space biquaternion Lorentz group transformations have constant biquaternion matrix elements that are analogous to those of the Lorenz group. (See eq. 10.7 for a Lorentz group transformation.)

5.5 Extracting the Symmetries and Particle Spectra

As stated earlier in section 5,2 we will directly describe the symmetry structure implied by the form of the biquaternion coordinate system while mindful of sections 5.3 and 5.4.

The Unified SuperStandard Theory developed the group structure from which the particle species were derived from a subset of Lorentz boost transformations. It used boosts with complex exponentiation similar to quaternion exponentiation in eq. 5.4. Complex boosts mapped a system at rest to a system in motion with a real energy and complex 3-momenta in general. Biquaternion boosts play a similar role.

The 4-dimensional representation of the Unified SuperStandard Theory complex coordinates was

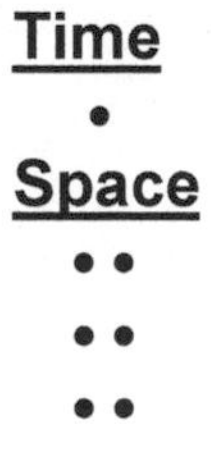

<u>Time</u>
•
<u>Space</u>
• •
• •
• •

Figure 5.2. Four-Dimensional subspace for Unified SuperStandard derivation of particle spectra with coordinates represented by • 's.

Following the same lines we now specify a biquaternion space restricted to that of Fig. 5.3 to define the relevant set of coordinates for determining particle symmetries and spectra.

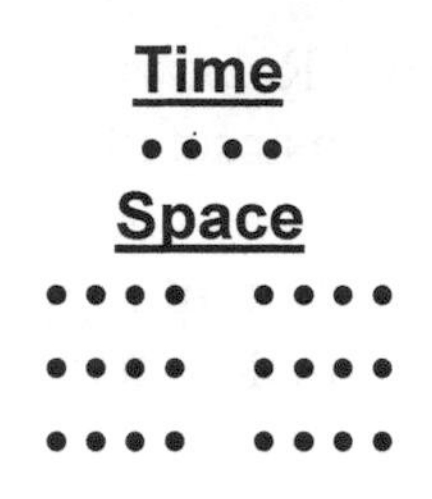

Figure 5.3. 4-dimensional biquaternion subspace for symmetries and particle spectra.

We expect its 14 complex coordinates will split into a 4-dimensional complex coordinate space which will support Complex Lorentz transformations, and a 10 complex coordinates internal symmetry space.

The mechanism for this symmetry breakdown may be due to vacuum energy effects in the biquaternion universe.

In the next chapter we analyze the symmetries of the 10 complex dimensional subspace and then augment these internal symmetries with Generation and Layer number symmetries. In MOST (chapters 13 and 14) we show the number symmetries are inherently part of the set of internal symmetries. We also show that Dark fermion sectors of the theory have spinors that occupy different parts of the overall 16 component spinors of the complex 8-dimensional space-time.

6. Symmetries of the Quaternion Unified SuperStandard Theory (QUeST)

The coordinate space picture of QUeST described in chapter 5 enables us to simply find the internal symmetries and particle spectra of QUeST. They will turn out to be those of the Unified SuperStandard Theory as presented earlier in chapter 2.

The QUeST coordinates with the space-time complex 4-vector separated from the internal symmetry coordinates is depicted in Fig. 6.1.

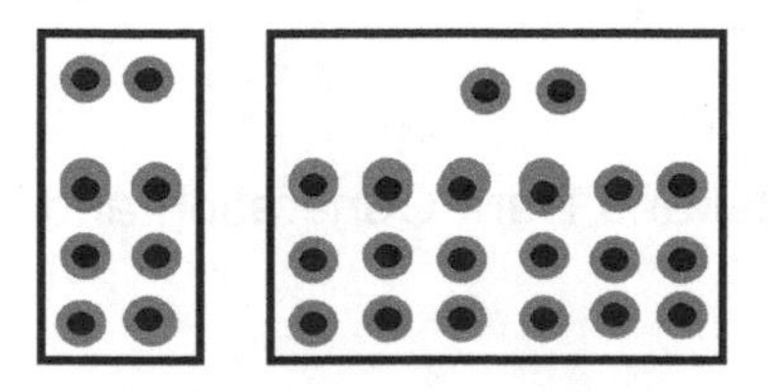

Figure 6.1. Eight real space-time coordinates separated from 20 real coordinates for internal symmetries.

The internal symmetry coordinates number 20 real coordinates or 10 complex coordinates. These coordinates serve to be coordinates[25] of the fundamental representations of each of the factors of

$$SU(2)\otimes U(1)\otimes SU(3)\otimes SU(2)\otimes U(1)\otimes SU(3)$$

---

[25] Simple counting of fundamental representation dimensions shows this to be true: 2 + 3 + 2 + 3 = 10 respectively. The set of 10 complex coordinates support transformations to a factorized form. The 10 complex coordinates can be transformed into a fundamental representation of the above factor groups.

The factorized internal symmetry emerges from another breakdown(s) which corresponds to the subgroups of the Lorentz group. They evidently follow from the structure of biquaternion Lorentz transformations.

6.1 Number Symmetries

The U(4) Generation and Layer groups do not appear in Fig. 6.1. Additional quaternion coordinates may be added to represent them. Four quaternions suffice to represent a U(4) for the normal sector and a U(4) for the Dark sector Generation groups, and a U(4) for the normal sector and a U(4) for the Dark sector Layer groups. Fig. 6.2 illustrates the coordinates for $U(4)^4$.

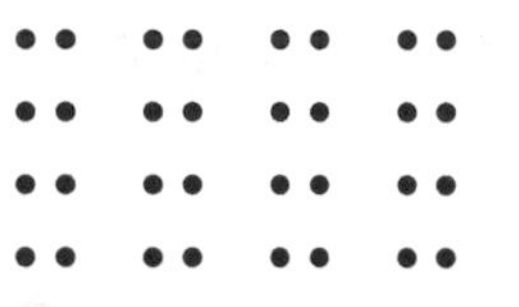

Figure 6.2. The four addional quaternions needed for normal and Dark Generation and Layer groups $U(4)^4$.

Later when we discuss MOST we will see that these additional quaternion coordinates arise naturally.

6.2 Layers

The QUeST figures above do not display the Internal Symmetry, Generation and Layer groups in four layers. The groups differ in the various layers. Their gauge fields are each flagged with a layer index. So the overall internal symmetry is the internal symmetry group of the Unified SuperStandard Theory:

$$[SU(2)\otimes U(1)\otimes SU(3)\otimes SU(2)\otimes U(1)\otimes SU(3)\otimes U(4)^4]^4$$

6.3 Fermion and Gauge Vector Boson Spectrums

The fermion and vector boson spectrums that emerge are those of the Unified SuperStandard Theory. They are displayed in Figs. 2.1 – 2.4.

6.4 Revised QUeST Axioms

AXIOMS

1. Biquaternion space exists and supports symmetry breaking to space-time and internal symmetries.
2. Quantum Field Theory supports fundamental particles that form a countable set. Each particle number operator is a generator in a particle interaction group.
3. All quantum field theory calculations are finite.
4. The Quantum Field Theory of particles can be defined in any curved space-time.
5. Each particle wave function has a functional defining the particle state with the functional in a set without a distance measure.

Figure 6.2 The axioms of QUeST.

7. Quantum Functionals, Wave Set, Projections Set

This chapter describes the basic method of the separation of quantum fields into quantum functionals and (fourier, …) waves, the rationale for the separation, the quantum functional set F, the wave set Φ, and wave-particle duality.[26]

7.1 Separation of Quantum Mechanical Wave functions into Waves and Quantum Functionals

In the past quantum fields have been thought to have the form

$$\psi_\zeta(x) = \Sigma\, (\,\dots\,)_\zeta \tag{7.1}$$

where ζ is a set of observables and quantum numbers, (…) is a wave function expression (often a fourier expansion) and Σ indicates a summation.

Previously we have proposed to separate Quantum fields (both quantum mechanical and quantum field theoretic) into quantum functionals and wave fields using a functional inner product. This mechanism also uses projections applied to the set of all possible "waves" to select relevant wave functions.

We begin by defining Φ to be the set of all possible "waves" meaning the set of all possible wave functions for all possible physical situations. Φ is an infinite set of generalized functions, members of which appear in quantum theory. We define a projection Π_ζ for wave function $(\,\dots\,)_\zeta$ of the ζth set of observables:

$$(\,\dots\,)_\zeta = \Pi_\zeta\, \Phi \tag{7.2}$$

Then the quantum functional[27] inner product yielding the quantum field is[28]

[26] Much of this chapter appears in Blaha (2018e).

$$\psi_\zeta(x) = f_\zeta(\Pi_\zeta\Phi\,) = (f_\zeta\,,\, \Pi_\zeta\Phi) \qquad (7.3)$$
$$= \Sigma\,(\,\ldots\,)_\zeta$$

where f_ζ is a quantum functional. We can combine the quantum functional and the projection

$$F_\zeta = f_\zeta\,\Pi_\zeta \qquad (7.4)$$

$$\psi_\zeta(x) = F_\zeta\,(\Phi) = (f_\zeta\Pi_\zeta\,,\, \Phi) = (F_\zeta\,,\, \Phi) \qquad (7.5)$$

since they appear together when defining a quantum field. Eq. 7.5 indicates the argument of F_ζ is the set of *generalized functions*[29] and thus we may call F_ζ a *generalized functional*[30].

Later we will see that projections are also used to define observations (reductions of wave packets).

We now turn to the more important task of explaining the reason for the separation of quantum fields into functionals and waves. Then we turn to the composition of the set of waves Φ and the set of quantum functionals.

7.2 Why Factorize Fermion Quantum Fields into a Functional and a Wave?

Quantum Mechanics and Quantum Field Theory both contain Quantum Entanglement. Quantum Entanglement phenomena are known to exhibit *instantaneous* (spooky) action-at-a-distance.[31] This seeming violation of the Theory of Relativity has sparked a continuing stream of papers and discussion. Proposals have been put forward to explain the apparent violation.

---

[27] Quantum functionals are functionals with additional features that will be specified later in this chapter.

[28] The indicated summation Σ often appears in quantum field expansions. Thus a functional inner product may contain an implied summation.

[29] Yu. V. Egorov, Russian Math. Surveys **45**, 1 (1990).

[30] Generalized functionals have not been considered in the literature to the best of the author's knowledge.

[31] A. Einstein, B. Podolsky, and N. Rosen, **47**, 777 (1935) plus an extensive recent literature.

We contend instantaneous action-at-a-distance is a fundamental quantum feature and thus ought to be based on a postulate of quantum theory. We account for instantaneous action-at-a-distance in quantum fields without a conflict with the Theory of Special Relativity by

1. Factorizing all quantum mechanical and quantum field theoretic wave functions into quantum functionals and waves;

2. Defining the set F of all quantum functionals to be in a 1:1 correspondence with all possible quantum wave functions;

3. Requiring that there is no distance measure in F so all points of space-time has instantaneous equal access to the functionals. We can view the set F as a "tensor product with space-time: " space-time⊗F".

Thus, as we shall see in a practical example in chapter 8, the instantaneity of quantum phenomena follows directly from the presence of the set F at all points of space-time. And there is no distance measure in F.

The reader may find a thorough discussion of particle functionals in Blaha (2019g) and (2018e) including the topic of particle cores, generated by functionals, which reside inside fermions and bosons: qubes and qubas.

7.3 General Form of Quantum Functionals

A *quantum functional* is a functional that obeys the algebraic rules specified in section 7.4. The set of quantum functionals F consists of all quantum functionals corresponding to possible quantum states. To the extent that all matter and energy is reducible to matter and energy particles, the elements of F are reducible (in principle) to combinations of sums and products of particle quantum functionals..

7.4 Quantum Functional Set Algebra

Quantum functionals support forms of addition (and subtraction) as well as multiplication. This section describes the algebra of quantum functionals. The form of the algebra defines the restricted set of quantum functionals.

7.4.1 Addition and Subtraction

Functionals of the set F′ can be added (subtracted) sensibly in some cases:

$$f_1 + f_2 = f_3 \tag{7.6}$$

meaning

$$f_3(f) = f_1(f) + f_2(f) \tag{7.7}$$

for some function f. However for quantum states a projection is involved. So for quantum states eq. 7.7 makes sense only if the projections Π_3, Π_1, and Π_2 are physically compatible. If so then

$$\begin{aligned} \psi_3(x) = f_3(\Pi_3\Phi\,) &= (f_1, \Pi_1\Phi) + (f_2, \Pi_2\Phi) \\ &= \psi_1(x) + \psi_2(x) \end{aligned} \tag{7.8}$$

Eq. 7.8 is the quantum interpretation of eq. 7.7. Eq. 7.8 shows quantum functionals support superposition of quantum wave functions.

7.4.2 Multiplication

The case of the multiplication of quantum functionals is slightly more complex. Note multiplication such as $f_3 = f_1f_2$ is not sensible since $f_1(f_2(f))$ is not well defined. The argument of f_1 must be a function and $f_2(f)$ is not a function.

We now consider multiplications involving quantum functionals.

7.4.2.1 Multiplication of Quantum Functionals by c-numbers

If we multiply a quantum functional by a c-number a

$$f_2 = af_1 \tag{7.9}$$

then the result is linear:

$$f_2(f) = af_1(f) \quad (7.10)$$

7.4.2.2 Linear Transformation of Quantum Functionals

If we linearly transform a set of quantum functionals by a matrix $[R_{ij}]$

$$f_i = R_{ij}f_j \quad (7.11)$$

then the result is

$$f_i(f) = R_{ij}f_j(f) \quad (7.12)$$

7.4.2.2 Multiplication of Quantum Functionals- Composite States

A form of multiplication of quantum functionals is possible in the case of composite quantum states. Consider two particles that form a quantum state. The multiplication could be specified as

.
$$f_3 = f_2f_1 \quad (7.13)$$

Eq. 7.13 symbolizes

$$f_3(f_3) = f_2(f_2)\ f_2(f_1) \quad (7.14)$$

where f_1 and f_2 are functions (waves) corresponding to two particles. In terms of quantum wave functions, we can express eq. 7.14 as

$$\psi_3(x_2, x_1) = f_3(\Pi_3\Phi\ ) = (f_1, \Pi_1\Phi)(f_2, \Pi_2\Phi) \quad (7.15)$$
$$= \psi_2(x_2)\psi_1(x_1)$$

where Π_i is the ith projection.

7.4.2.2 Functionals Combined by Interactions

Later we will consider lagrangians with interactions that are skeletonized to consist of terms containing products of quantum functionals. For example, consider a quantum functional interaction for

$$e+e- \rightarrow \gamma$$

electron – positron annihilation to a photon. The quantum functional expression could be expressed as

$$f_\gamma f_{e+} f_{e-} \tag{7.16}$$

In perturbation theory eq. 7.16 would have the form

$$\int d^4x\ \psi_\gamma(x)\psi_{e+}(x)\psi_{e-}(x)\ldots \tag{7.17}$$

where … represents exponential momentum factors.

Thus particle transitions can be represented by products of quantum functionals.

7.5 Wave Set Φ

The wave set consists of all generalized functions[32] that could serve as part of quantum wave functions or field operators. Thus it is infinite dimensional. It is also more general than Hilbert spaces. The members of the set are effectively localized to a point that is accessible to all space-time points.

The wave functions, which are often a form of fourier expansion in the free case, are transformed by quantum functionals into quantum wave functions and field operators.

7.6 Projection Set Π

The set of projections Π is an infinite set whose members are in 1:1 relation with the set of quantum functionals and in 1:1 relation with the set of all quantum wave functions and field operators.

Members of this set have the multiplication rule:

$$\Pi_i\Pi_j = \delta_{ij}\Pi_i \tag{7.18}$$

where δ_{ij} is the Kronecker delta symbol.

[32] Egorov *op. cit.*

The elements of the set operate on Φ to select the wave function which a quantum functional transforms into a quantum wave function or field operator. Eq. 7.8 is a simple example.

The set of all projections is localized to a point that is instantaneously accessible at every space-time point. Elements of the set and sums of elements of the set can be used to specify measurements (reductions of quantum states).

7.7 Particle-Wave Duality Mathematically Realized

At the beginning of Quantum Theory in 1924, De Broglie postulated wave-particle duality and made progress in understanding quantum phenomena. Wave-particle duality was subsequently 'abandoned' in favor of the Quantum Mechanics of Heisenberg and Schrődinger.

The quantum field factorization that we propose provides a mathematical formulation of particles that separates the particle part (the quantum functional) from the wave part (the wave). Thus we have realized wave-particle duality for the purpose of understanding instantaneous effects of quantum entanglement.

In our formulation, quantum fields $\chi(\mathbf{x}, t)$ appear in the SuperStandard lagrangian. *Then* perturbation theory calculations of phenomena are made using free field wave expansions, denoted $(\mathbf{x}, t)$, for all fermions and bosons. Thus we achieve a quantum theory (non-deterministic) and yet have wave-particle duality embedded within it (unlike De Broglie-Bohm theory.)

7.7.1 Visualization of a Particle

We can visualize an elementary particle located at point x as composed of two components: a quantum functional **f** and a wave denoted **x**:[33]

[33] See Blaha (2019g) and (2018e).

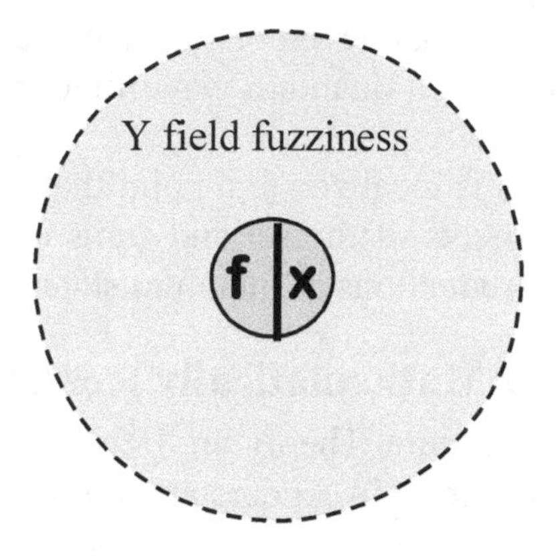

Figure 7.1 Symbolic view of a free particle having particle functional and fourier expansion parts.

The central disk represents the core of the particle which is located at x but 'smeared' by a cloud of Y particles generated from Two-Tier Quantization. Thus point particles do not exist in the full Unified SuperStandard Theory. Thus infinities are not encountered in the calculation of any diagram in its perturbation theory.

The central disk represents a Qube functional for fermions and a Quba functional for bosons. (See Blaha (2019g) and (2018e).) As described earlier Qubes embody spin ½ and Qubas embody integer spin. Qubes have a bare mass m_0 and Qubas are massless. The bare mass of Qubes reflects the fact that fermions have spin, and thus an intrinsic energy, while Qubas can carry energy and thus spin while in motion but elementary "bosons act only as 'delivery agents' for spin.

7.8. Quantum Functionals Implement Plato's Theory of Ideals

Quantum functionals implement Plato's Theory of Ideals and Reality. He viewed ideals as the abstract representation of things in nature that embodied the nature of things. He viewed reality as the items in nature and thought that we encounter. Thus there are dogs in nature. "Dogness" is the ideal represented by dogs. Similarly, love between two people is reality. Love, the Ideal, embodies the concept of individual instances of real love.

In our theory, quantum functionals are the equivalent of an Ideal in that they represent a thing in itself —a mathematical specification. Reality is the quantum wave functions (and fields) that represent specific things.

7.9 Quantum Functionals—A Buddhist-like Unification of the Physical Cosmos

In Buddhist and some other Eastern Cosmologies and religions the Cosmos is viewed as an interconnected unity with all parts interconnected.

The point set (universe) of quantum functionals, which exists at all points of space-time, which instantaneously interconnects all parts of the web of reality, which establishes the reality of events in the universe, and which is a unity of all possible states, gives a reality to the abstract Cosmology of the East.

In a sense the universe is One—quantum mechanically united.

8. Functionals Implement Quantum Entanglement

In 1935 Einstein, Podolsky, and Rosen[34] (EPR) proposed a definition of Physical Reality and then proceeded to consider a quantum state composed of two separated quantum systems. In this chapter we will propose an extension of their definition of Physical Reality, and then show, by counterexample, that the clarification is needed.

Then we will consider their separated quantum systems example within the framework of our quantum functional formalism. We will show our formalism eliminates instantaneous quantum action-at-a-distance. (No spookiness.)

8.1 Definition of Physical Reality

EPR suggested the definition of Physical Reality:

EPR Criterion for Physical Reality

If, without in any way disturbing a system, we can predict with certainty (i.e., with probability equal to unity) the value of a physical quantity, then there exists an element of physical reality corresponding to this physical quantity.

EPR suggested there might be other ways of defining physical reality in a manner consistent with Physical Theory. It appears that a more comprehensive definition of physical reality is required to justify a correspondence with Physical Theory.

The criterion for a correspondence between physical reality and a physical theory must include the requirement that a physical theory's complete set of observables must correspond to physical reality. Predicting with certainty the value of one physical

[34] Einstein *op. cit.*

quantity leaves open the question of other possible physical quantities. In the case of conventional simple quantum theory there are two observables Q and P with commutation relation [Q, P] = i. Within this framework the EPR definition appears acceptable. However one should not hang the definition of physical reality on one possible quantum mechanics formalism. Physical reality must be independent of all valid formulations. A simple counter example shows that the EPR definition is not sufficient in the case of an alternate quantum theory.

Counter Example

Consider the formalism of Blaha (2016f) which was used to create a larger quantum theory that encompassed both quantum physics and classical physics as extremes. The theory supports both classical and quantum physics and an intermediate range between them. In this theory (called PseudoQuantum Mechanics) the range from quantum to classical physics is specified by an angle θ.

The theory had two commuting variables x_1 and p_1, which we augment with two new variables x_2 and p_2, defined by

$$x_i = (m\omega/\hbar)^{-1/2} Q_i \qquad (8.1)$$
$$p_i = (m\omega\hbar)^{1/2} P_i$$

for i, j = 1, 2 where

$$P_2 = -i\, d/dQ_1 \qquad (8.2)$$
$$Q_2 = i\, d/dP_1$$

with the commutation relations:

$$[Q_i, P_j] = i(1 - \delta_{ij}) \qquad (8.3)$$

for i, j = 1, 2.

Next we define raising and lowering operators

$$a_i = 2^{-1/2}(Q_i + iP_i) \qquad (8.4)$$
$$a_i^\dagger = 2^{-1/2}(Q_i - iP_i)$$
$$Q_i = (a_i + a_i^\dagger)/\sqrt{2}$$
$$P_i = (a_i - a_i^\dagger)/(\sqrt{2}i)$$

with

$$[a_i, a_j^\dagger] = (1 - \delta_{ij}) \quad (8.5)$$
$$[a_i, a_j] = 0$$
$$[a_i^\dagger, a_j^\dagger] = 0$$

for i, j = 1, 2.

We now define an alternate set of raising and lowering operators that will use an angle θ to provide a continuous transition from classical to quantum (and vice versa)

$$b_1 = Q_1\cos\theta + iP_2\sin\theta \quad (8.6)$$
$$b_2 = -Q_2\sin\theta + iP_1\cos\theta$$

$$b_1^\dagger = Q_1\cos\theta - iP_2\sin\theta \quad (8.7)$$
$$b_2^\dagger = -Q_2\sin\theta - iP_1\cos\theta$$

Their commutation relations are

$$[b_1, b_1^\dagger] = \sin(2\theta) \quad (8.8a)$$
$$[b_2, b_2^\dagger] = -\sin(2\theta)$$
$$[b_1, b_2^\dagger] = [b_2, b_1^\dagger] = 0$$
$$[b_1, b_2] = [b_1^\dagger, b_2^\dagger] = 0$$

The PseudoQuantum Hamiltonian is

$$\hat{H} = p_1p_2/m + m\omega^2x_1x_2 \quad (8.8b)$$
$$= \tfrac{1}{2}\omega(\{a_1, a_2^\dagger\} + \{a_2, a_1^\dagger\})$$
$$= \omega(P_1P_2 + Q_1Q_2)$$

Note: all physical quantities commute for θ = 0 by eq. 8.8a. But they are quantum for θ = π/4. Thus this theory encompasses both classical and quantum theory. Blaha (2016f) develops a harmonic oscillator example (and other examples) showing both its classical and quantum limits as well as intermediate oscillator states.

The example illustrates the point that predicting the value of one physical quantity with certainty does not make it physically real. All physical quantities of a system must also have commutation relations in agreement with Physical Reality. The *above counterexample illustrates this point by defining a sharp physical quantity with variable commutation relations. The choice of the parameter θ leads to a quantum theory or a classical theory or an intermediate theory.*

A more satisfactory criterion for Physical Reality is:

New Criterion for Physical Reality

If, without in any way disturbing a system, we can predict with certainty (i.e., with probability equal to unity) the value of a physical quantity and determine the commutation relations of all observables in the Physical Theory, then there exists an element of physical reality corresponding to this physical quantity and to the enveloping Physical Theory.

The commutation relations of the physical quantity are critical to the reality question. The question of physical reality is dependent on the complete formulation. Thus the question of physical reality depends on more than the value of each physical quantity. The reality question requires a total specification of commutation relations and eigenvalues.

8.2 Quantum Entanglement and Action-at-a-Distance

EPR considered the quantum entanglement of two systems and showed that instantaneous action-at-a-distance (spookiness) resulted. In this section we will show that our quantum functional formalism, which generalizes quantum theory, eliminates the problem of instantaneous action-at-a-distance.[35] We will show the solution provided by quantum functionals using the same example as EPR.

The key feature of quantum functionals is their ubiquitous presence at every point of space-time. In a multi-system quantum state all functionals are directly, instantaneously linked no matter what the separation of the constituent systems. When a reduction of the state of one system occurs due to a measurement, all other systems are instantly updated since the space-time separation of the individual systems is not

[35] This section presents the solution for quantum spookiness that we proposed in Blaha (2019g) and (2018e).

relevant. The linkage of all quantum functionals is relevant. A reduction of one system immediately impacts the other related systems.

8.3 The EPR Two System State Example

EPR considered a state consisting of two systems that might become separated spatially. We can represent the state as

$$\Psi = \Sigma_n \psi_{1n}(x_1)\psi_{2n}(x_2) \tag{8.9}$$

We can represent a measurement (reduction of state) with a projection Π_{1a} of system "1" to a state ψ_{1a} with

$$\psi_{1a} = \delta_{ab}\, \Pi_a\, \psi_{1b} \tag{8.10}$$

Then

$$\Psi_{\text{projected}} = \Pi_{1a}\, \Sigma_n \psi_{1n}\psi_{2n} = \psi_{1a}(x_1)\psi_{2a}(x_2) \tag{8.11}$$

The effect of the measurement of system "1" is *instantaneous* of system "2" because the quantum functionals f_{1n} and f_{2n}, and the projections Π_{1n} and Π_{2n} of both systems are not separated by distance. By eq. 7.3

$$\psi_{1n}(x) = f_{1xn}(\Pi_{1xn}\Phi\,) = (f_{1xn}, \Pi_{1xn}\Phi) \tag{8.12}$$
$$\psi_{2n}(y) = f_{2ny}(\Pi_{2yn}\Phi\,) = (f_{2yn}\,, \Pi_{2yn}\Phi) \tag{8.13}$$

with $x = x_1$ and $y = x_2$. *The quantum functional and the projection select the wave and its coordinate parameterization. The coordinates in the wave are merely place holders.*

Therefore the relative distance between the coordinates x_1 and x_2 is not relevant for the change of state of system "2". The quantum functionals and projections give the instantaneity of the change in ψ_{2a} upon the measurement of system "1".

The EPR Spookiness is resolved by quantum functionals. There is no conflict with the Theory of Special Relativity.

Particle Functional Theory (PFT)

9. Particle Functional Theory

In previous chapters we have seen that we have a point space with no distance measure containing the set of all quantum functionals. This space enables us to define quantum functionals that ultimately support instantaneous Quantum Entanglement effects (Chapter 8). We saw quantum fields can be defined as an inner product of quantum functionals and wave fields.

Based on the set of quantum functionals defined earlier we can construct quantum functionals for all the fundamental fermion and boson fields. We can define quantum field interactions. Quantum functionals, combined with wave expressions in coordinate space can then be used in perturbation theory to calculate probabilities.

While we define quantum functionals for fundamental elementary particles, combinations of particles such as bound states and molecules as well as clumps of matter will have combinations of quantum functionals corresponding to their aggregated constituents, Ultimately all matter in the universe has quantum functionals describing them. Since the number of particles in the universe is finite but very large at any instant of time, the number of quantum functionals in the universe is also finite (modulo coordinates) but is large but countable at any instant of time. As time progresses the number of functionals varies due to particle creation and annihilation.

The universe that emerges from Particle Functional Theory consists of the direct product of space-time with the space of internal symmetries. And every point in space-time also interfaces with the functional point space. The functional point space is one entity that is ubiquitously contiguous with each space-time point.

Thus the universe achieves a unity through Particle Functional Theory (PFT). Simultaneous Quantum Entanglement is a consequence of this theory. PFT removes the spookiness of entanglement since all particle symmetries and spins are resident at one point. Spooky Quantum Entanglement is not a disease. It is an indication of a deep feature of reality.

The indices of a particle quantum functional are:

1. Internal Symmetry Eigenvalues
2. Total Spin
3. "z-component" of spin
4. Mass (after an inner product to create a particle wave function)
5. Particle position and/or momentum

The general form of the inner product creating a quantum mechanical wave function and a quantum field theoretic wave function is

$$\psi_\zeta(x) = f_\zeta(\Pi_\zeta\Phi) = (f_\zeta , \Pi_\zeta\Phi) \tag{7.3}$$

where ζ specifies the indices, Π_ζ is the corresponding projection, and Φ is the set of all waves.

10. Particle Cores, The Four Fermion Species, and Their Second Quantization

In this chapter we outline the origin of the fermion spectrum in the Unified SuperStandard Theories.

10.1 The Logic Core of Fundamental Fermions and Bosons

Blaha (2017f) opened the possibility that fermions (and bosons) might have a core functional that embodies logic in the form of internal symmetries and spin as well as bare masses in the case of fermions. Subsequently we expanded our discussion to better describe the functionals set within which the core functionals of each type of fundamental particle reside. We also defined functionals of various spins: 0, ½, 1, and 2. We saw that the cores of spin ½ fermion functionals (that we call *qubes* in analogy with qubits) have four varieties as well as internal symmetries.[36] Fermion functionals also have a bare mass denoted m_0.[37]

Bosons have cores as well that are boson functionals with integer spin and internal symmetries. We call a boson core a *quba*[38] in analogy with the fermion functionals name of qubes. Boson functionals are massless. Bosons may acquire masses through interactions.

The most significant rationale for logic cores for particles is that the formalism, based on the set of all particle functionals, leads to an explanation of the 'spooky' action at a distance of Quantum Entanglement—a subject of much discussion.

---

[36] Each of the four varieties (which we call species) can be separated into left-handed and right-handed functionals.

[37] Although we say a fermion functional has a bare mass, the functional only acquires the bare mass after it is folded into a fourier coordinate expression to become a quantum field. Since functionals exist in a functional space with no distance measure, functional mass in itself is not meaningful.

[38] We use 'quba' simply because of its similarity to 'qube'. The leading 'b' signifies its bosonic use. We pronounce 'quba' as 'bub' with a silent 'e.' The word 'quba', itself, is the name of a Bantu language spoken by the Bubi people of Bioko Island in Equatorial Guinea.

10.2 The Logic Building Block of Fermions – Qube Cores

If we consider all possible 'things' that might constitute a fundamental building block for a fundamental fermion theory they are all, at best, *ad hoc* and raise questions of their necessity, as well as whether they are composed of yet a more fundamental substructure.

There is only one choice of building block that avoids these issues – a logic unit or qubit. A qubit is a fundamental entity that is a complex form of computer bit. A bit (and thus a qubit) is known to have an energy or equivalently a mass, and has no constituents of a more primitive form.[39] We call a unit of logic that forms the core of a particle a *qube*.[40] It exists as the core of a particle although it is located in a space of functionals. In itself it has no *independent* material existence or space-time coordinates. A qube is a quantum functional that acquires features such as coordinates, to become an elementary particle. Quantum functionals are functions whose arguments are functions—not variables.

We define a qube as a fermion field theory quantum functional. Later we describe other physical features that "cloak" qubes with properties and interactions making them into fundamental fermion quantum fields.

10.3 Mass of a Qube

Recent experiments have shown that a logical value of a qubit has an energy.associated with it. One bit of information has about 3 x 10^{-21} joules of energy[41] or a rest mass, m_0, or about 0.02 eV using $m_0 = E/c^2$. This result was confirmed by E. Lutz et al.[42] who showed that there is a minimum amount of heat produced per bit of erased data. This minimal heat is called the *Landauer*[43] *limit*. The equivalent mass we

[39] A qube is a physical manifestation of a logical value. The relation of a qube to a logical value is analogous to the relation of a penciled point placed on paper to the concept of a point as a primitive in geometry.
[40] In the Blaha (2017f) we called qubes iotas. However, since the name iota was previously used as a particle name many years ago it seemed reasonable to use a different name. We chose the name 'qube' for self-evident reasons. *'Qube' is pronounced 'cube.'*
[41] E. Muneyuki et al, *Nature Physics*, DOI: 10.1038/NPHYS1821.
[42] E. Lutz et al, Nature **483** (7388): 187–190,10.1038/nature10872, (2012).
[43] R. Landauer, "Irreversibility and heat generation in the computing process", IBM Journal of Research and Development **5** (3): 183–191, (1961).

will call the *Landauer mass* and denote it as m_0. We will assume that a fundamental Landauer mass exists in our discussions although the precise value of the mass will not be used since we may expect all physical particle masses to be renormalized to different values when interactions are taken into account.

We will assume all fermions contain a qube within them. A qube is assumed to "have" mass m_0. Although we say a fermion functional has a bare mass, the functional only "acquires" the bare mass after it is folded into a fourier coordinate expression to become a quantum field. Since functionals exist in a functional set with no distance measure, functional mass in itself is not meaningful. The masses of fermions are modified to their known values by interactions.

It is intriguing that the mass of the electron neutrino has been measured in a variety of experiments and found to be within an order of magnitude or so larger than our estimate of the Landauer mass (as we would expect since particles acquire a 'cloud of virtual particles' due to interactions.) This 'cloud' can be expected to increase its mass above the Landauer mass. Since neutrinos only have the weak interaction it is not surprising that the increase due to interactions should not be large. The Mainz Neutrino Mass Experiment, for example, estimates the electron neutrino mass to be less than 2 eV. New experiments suggest an upper bound of 1.1 eV (KATRIN experiment).

A number of astronomical studies have also generated estimates of neutrino masses. In July 2010 the 3-D MegaZ DR7 galaxy survey found a limit for the combined mass of the three neutrino varieties to be less than 0.28 eV.[44] A smaller upper bound for the sum of neutrino masses, 0.23 eV, was found in March 2013 by the Planck collaboration,[45] In February 2014 a new estimate of the sum was found to be 0.320 ± 0.081 eV due to discrepancies between the Planck's measurements of the Cosmic Microwave Background, and other predictions, combined with the assumption that neutrinos are the cause of weaker gravitational lensing than implied by massless neutrinos.[46]

---

[44] S. Thomas et al, "Upper Bound of 0.28 eV on Neutrino Masses from the Largest Photometric Redshift Survey", Physical Review Letters **105**: 031301 (2010).

[45] Planck Collaboration, arXiv:1303.5076 (2013).

[46] R. A. Battye et al, "Evidence for Massive Neutrinos from Cosmic Microwave Background and Lensing Observations", Phys. Rev. Lett. **112,** 051303 (2014).

Thus the experimentally measured values of neutrino masses are consistent with the qube Landauer mass estimate of 0.02 eV given above. We thus assume that a fermion particle consists of a qube with a certain mass[47] that is renormalized, together with other features. These features emerge in the derivation of the complete theory.[48]

We view Reality as ultimately a representation (or painting) of logic values evolving through interactions in time and space.[49]

10.4 Qube Spin

The spin of a qube for a fundamental fermion is assumed to be spin ½. Qubes are solely a building block of fermions.

10.5 Qubes as Fermion Field Functionals

At this point qubes have an insubstantial appearance with only the attributes of internal symmetry, spin and mass. We have suggested that they can be mathematically represented as fermion field quantum functionals[50] and used to develop the structure of the fermion spectrum and The Unified SuperStandard Theory.

We see that the Standard Model interactions and features such as Quantum Entanglement *require* the use of a functional formalism for particle fields.

This new deeper formulation supports the theory presented in the Blaha (2017f). It adds a new level of depth that extends and clarifies the theory presented there.

We now define a canonical quantum functional approach to creating a simple Dirac fermion quantum field from a qube and a fourier quantum expression for the space-time part of a free fermion quantum field. Initially we symbolize a qube for a

---

[47] Leibniz first proposed the idea of logic 'particles' which he called monads. Our definition of a logic 'particle' does not include (or exclude) the presence of a spiritual part which was part of the definition of Leibniz's monads.

[48] A recent experiment claims to separate the spin part (which we identify as a logical value later) of a molecule from the rest of the molecule.

[49] Those who might suggest matter is substantial, and logic values are not, should remember that matter would be completely insubstantial if there were no forces in nature. Neutrinos which are close to insubstantial would be completely insubstantial if there were no weak interactions.

[50] Functionals are a mathematical primitive of our theory. They have been used extensively by Feynman and others in quantum theories.

fermion with a set of internal symmetry eigenvalues ξ, and spin s as $f_{\xi s}$.[51] We begin by defining a coordinate space Dirac fourier quantum expansion for internal symmetry eigenvalues ξ as

$$(s, x, t)_\xi = N(p)[\, b_\xi b(p, s)u(p, s)e^{-ip\cdot x} + d_\xi^\dagger d^\dagger(p, s)v(p, s)e^{+ip\cdot x}]$$

where N(p) is a normalization factor, u and v are functions of spin and momentum, and b and $d^\dagger$ are creation/annihilation operators.

A Dirac quantum wave function can be defined as an inner product of a qube functional and a coordinate space fourier quantum expansion. For example

$$\psi_\xi(x) = (f_\xi, (s, x, t)) = \sum_{\pm s} \int d^3p N(p)[b_\xi b(p, s)u(p, s)e^{-ip\cdot x} + d_\xi^\dagger d^\dagger(p, s)v(p, s)e^{+ip\cdot x}] \quad (10.1)$$

where we use a functional inner product formalism in the manner of Riesz (1955) and others. Eq. 10.1 can be viewed as an inner product but it is more correct to view it as a projection by the functional on the set of all possible coordinate space expansions. We note that the b_ξ and $d_\xi^\dagger$ coefficients transform to internal symmetry matrices in perturbation theory computations.

A functional inner product yields a numeric value. In the present case, it yields a numeric (possibly quantum) function. In general an inner product of a functional f with a variable function g is expressed as

$$G(x) = (f, g(x))$$

For each value of x, G(x) has one numeric value modulo quantum smearing.

10.6 Boson Elementary Particles – Quba Cores

In defining qubes above we have only considered the fermion case. For reasons that will be apparent later there is also a need for boson core functionals called qubas. We can define a corresponding boson functional quba for each type of boson. We will

[51] And possibly quantum coordinates x and p.

designate a boson quantum functional as $b_{\xi s}$ where s specifies the spin and ξ specifies the internal symmetry eigenvalues. Every boson contains a boson functional core within it. For consistency we called a boson functional a *quba*. A quba has the spin of the elementary boson within which it resides. It has the appropriate internal symmetry eigenvalues. It has zero mass since bosons are typically massless prior to symmetry breaking effects. The functional content embodied in each type of elementary particle is summarized in Table 10.1. (A functional also has space and/or momentum labels.)

PARTICLE TYPE	CORE	MASS	SPIN	INTERNAL SYMMETRY INDICES
Fermion	qube	m_0	½	ξ
Scalar Boson	quba	0	0	ξ
Vector Boson	quba	0	1	ξ
Graviton	quba	0	2	none

Table 10.1 Core functionals within the various types of fundamental elementary particles. Dependence on coordinates and momenta is not displayed.

10.7 Matrix Representation of Complex Lorentz Group L_C Boosts

The remainder of this chapter is based on the Complex Lorentz group which will be seen to have a primary role in defining the structure of the fermion spectrum.

We begin with Complex Lorentz Group (L_C) boosts because they will be crucial in the determination of the equations of motion of various types of spin ½ particles. An L_C boost can be expressed in the form[52]

$$\Lambda_C(\mathbf{v_c}) = \exp[i\omega\hat{\mathbf{w}}\cdot\mathbf{K}] \tag{10.2}$$

where

$$\omega = (\omega_r^{\,2} - \omega_i^{\,2} + 2i\omega_r\omega_i\,\hat{\mathbf{u}}_r\cdot\hat{\mathbf{u}}_i)^{1/2} \tag{10.3}$$

and

$$\hat{\mathbf{w}} = (\omega_r\hat{\mathbf{u}}_r + i\omega_i\hat{\mathbf{u}}_i)/\omega \tag{10.4}$$

Since $\hat{\mathbf{u}}_r\cdot\hat{\mathbf{u}}_r = 1 = \hat{\mathbf{u}}_i\cdot\hat{\mathbf{u}}_i$

[52] See Blaha (2019g) and (2018e) for more detail.

$$\hat{\mathbf{w}}\cdot\hat{\mathbf{w}} = 1 \tag{10.5}$$

and the complex relative velocity is

$$\mathbf{v}_c = \hat{\mathbf{w}} \tanh(\omega) \tag{10.6}$$

We now analytically continue to complex ω and complex unit vectors $\hat{\mathbf{w}}$. The resulting complex generalization will be the matrix form of proper L_C boosts:

$$\Lambda_C(\mathbf{v}_c) = \exp[i\omega\hat{\mathbf{w}}\cdot\mathbf{K}] \equiv \Lambda_C(\omega, \hat{\mathbf{w}})$$

$$= \begin{bmatrix} \cosh(\omega) & -\sinh(\omega)\hat{w}_x & -\sinh(\omega)\hat{w}_y & -\sinh(\omega)\hat{w}_z \\ -\sinh(\omega)\hat{w}_x & 1+(\cosh(\omega)-1)\hat{w}_x^2 & (\cosh(\omega)-1)\hat{w}_x\hat{w}_y & (\cosh(\omega)-1)\hat{w}_x\hat{w}_z \\ -\sinh(\omega)\hat{w}_y & (\cosh(\omega)-1)\hat{w}_x\hat{w}_y & 1+(\cosh(\omega)-1)\hat{w}_y^2 & (\cosh(\omega)-1)\hat{w}_y\hat{w} \\ -\sinh(\omega)\hat{w}_z & (\cosh(\omega)-1)\hat{w}_x\hat{w}_z & (\cosh(\omega)-1)\hat{w}_y\hat{w}_z & 1+(\cosh(\omega)-1)\hat{w}_z^2 \end{bmatrix} \tag{10.7}$$

Since analytic continuations are unique, the above form for $\Lambda_C(\mathbf{v}_c)$ is well-defined and unique. It spans the complete set of proper L_C boosts.

10.8 First Step Towards The Unified SuperStandard Theory

Thus we have derived a set of four fermion species that corresponds to the known fermions of one fermion generation from the Complex Lorentz Group.[53] We can derive the one generation form of the model in detail from Complex Lorentz group features. Then we can derive the four fermion generation form of the model based on the U(4) Generation group.

The overall pattern that begins to emerge from the developments in this chapter divides particles and interactions into two categories (as seen in Nature):

---

[53] Complex Lorentz group boosts lead to tachyons.

Particles with real 4-Momenta	Complexons (Complex 3-Momenta)
Leptons	Color quarks
SU(2)⊗U(1) Vector Bosons	Color SU(3) gluons
Higgs Particles	Possibly Higgs Particles

Basically the leptons, the SU(2)⊗U(1) Vector Bosons and the set of Higgs particles are be primarily based on the Left-handed boosts. These particles have real energies and momenta although some are "normal" and some are tachyons.

Another category of particles, complexons, emerges from our study of L_C. These particles have real energies and complex 3-momenta. In perturbation theory the loop integrations of loops of these particles would consist of a 7-fold integration over energy and complex 3-momenta with corresponding 7-fold delta functions to enforce energy-momentum conservation. As pointed out earlier the complex 3-momenta of these types of fermions have an SU(3) symmetry that it is natural to generalize to local color SU(3). (The other category of fermions, leptons. lack global SU(3) symmetry.) Thus we see the beginnings of the structure of the Unified SuperStandard Model in this chapter on spin ½ particles.

10.9 Dirac-like Equations of Matter from 4-Valued Logic

In our derivation every truly fundamental particle of matter, whether quark or lepton, has spin ½. We have seen in chapter 10 of Blaha (2011c) that the basic algebra of Operator Logic eigenvalue operators, and that of its raising and lowering operators, is the same as the algebra of creation and annihilation operators for free spin ½ particles. Our goal is to build our theory on the scaffolding of Operator Logic. We view a fermion particle as a qube core which is dressed in spatial coordinates (and internal symmetries):[54]

$$\text{Qube core} \circ \text{coordinates} \rightarrow \text{fermion particle} \qquad (10.8)$$

[54] The symbol ○ indicates an inner product of a functional with a function argument.

The creation and annihilation operators b(p,s) and $d^{\dagger}$(p,s) (and their hermitean conjugates $b^{\dagger}$(p,s) and d(p,s)) are mathematically similar to the raising and lowering operators of Operator (Matrix) Logic. They satisfy the anticommutation relations

$$\{b(q,s),\, b^{\dagger}(p,s')\} = \delta_{ss'}\delta^3(\mathbf{q} - \mathbf{p}) \qquad (10.9)$$
$$\{d(q,s),\, d^{\dagger}(p,s')\} = \delta_{ss'}\delta^3(\mathbf{q} - \mathbf{p})$$

Thus we see spin ½ particle wave functions originating from the Dirac-like spinors, and raising and lowering operators of the spinor formulation of Operator Logic.

When particles interact the quantum field theory interaction terms use fermion creation operators, b(q,s) and $d^{\dagger}$(q,s), *and annihilation operators,* $b^{\dagger}$(p,s′) and d(q,s), *to implement the transformations between the Qubes of the interacting particles. Thus the mathematics of the embedded Qubes' logic values is automatically implemented within quantum field theoretic calculations.*

An interesting point that emerges from this discussion is the nature of spin ½ particle states such as

$$|p, s\rangle = b^{\dagger}(p, s)|0\rangle \qquad (10.10)$$

This state is interpreted as a one particle state. It also has an analogous interpretation in Operator Logic as creating a one term universe of discourse – a construct which is in part linguistic and in part logic. Thus particles are embodiments of Logic values. Particle interactions change the logic values of the initial particles to those of the emergent particles. All in all, our universe can be viewed as an extraordinarily intricate logic machine. Serendipitously we are now seeing the use of particles to create quantum computers, which, in a sense, is bringing us full circle. Particles are Logic; Logic machines emerge from particle interactions.

10.10 Why Second Quantization of Fields?

One might have argued that the fermion field types that we have found could be treated as ordinary c-number fields and not be second quantized. However, particles are discrete entities that can be enumerated with integers. Second quantization implements the discrete particle concept in the most direct way and thus by Leibniz's Principle, as well as Ockham's Razor, second quantization is the best solution to obtain particle discreteness.

Quantum Theory is required by the discreteness of particles.

10.11 Why lagrangians? For dynamic evolution

Lagrangians naturally emerge as the 'preferred' formalism for quantum field theory due to their intimate relation with the energy-momentum tensor (particularly the Hamiltonian) that provides the generators of time evolution and of spatial translation.

10.12 Functional Expression for Each of the four Species of Fermions

We have derived the four species of fermions. We have used a 'conventional' notation for quantum fields. In this section we will define these quantum fields as "inner products" of functionals and fourier coordinate expansions.[55]

Dirac Quantum Field:

$$\psi(x) = ({}_1f, \text{Dirac\_fourier\_expansion})$$

Tachyon Quantum Field:

$$\psi_T(x) = ({}_2f, \text{Tachyon\_fourier\_expansion})$$

Complexon Quantum Field:

$$\psi_C(x) = ({}_3f, \text{Complexon\_fourier\_expansion})$$

Complexon Tachyon Quantum Field:

$$\psi_{CT}(x) = ({}_4f, \text{Tachyon\_Complexon\_fourier\_expansion})$$

The digit prefixes of ${}_kf$ for k = 1, 2, 3, 4 distinguish the functionals for each species.

In addition we can decompose the above quantum fields into left-handed and right-handed fields. The left-handed functional representations are:

[55] Although we note that quantum fields are generated by functional projections.

Left Dirac Quantum Field:

$$\psi_L(x) = ({}_{1L}f, \text{left-handed\_Dirac\_fourier\_expansion})$$

Left Tachyon Quantum Field:

$$\psi_{TL}(x) = ({}_{2L}f, \text{left-handed\_Tachyon\_fourier\_expansion})$$

Left Complexon Quantum Field:

$$\psi_{CL}(x) = ({}_{3L}f, \text{left-handed\_Complexon\_fourier\_expansion})$$

Left Complexon Tachyon Quantum Field:

$$\psi_{CTL}(x) = ({}_{4L}f, \text{left-handed\_Tachyon\_Complexon\_fourier\_expansion})$$

The right-handed cases have analogous forms.

11. Quantum Functional Lagrangian Skeletons

Quantum field theory calculations are almost always performed in perturbation theory. Perturbation theory expansions[56] use vacuum expectation values of time ordered products of pairs of quantum fields. Since quantum fields, in our functional wave-particle formulation, are the result of inner products of functionals (particle cores) and waves (fourier wave expansions), the form of quantum field vacuum expectation values is the same as usually found.

Therefore our new deeper level of our understanding of particle structure does not change perturbation theory. However it does account for the instantaneity of effects in separated parts of a quantum entangled process.

11.1 Skeleton Functional Lagrangians

If we could imagine a 'snapshot' of the universe[57] at one instant of time we could presumably enumerate all the functionals of the universe's particles. Then succeeding snapshots would show an ebb and flow of functionals as time progresses. This thought brings us to the important issue of the transformations of particle functionals in particle interactions. The simplest statement that one could make about functional transformations is that they are created and annihilated according to the interaction terms of the skeletonized Unified SuperStandard Theory lagrangian (excluding quadratic terms which do not transform functionals.)

We skeletonize a lagrangian density by deleting all quadratic terms and replacing all particle fields by their corresponding functionals.[58] For example the lagrangian

---

[56]See chapter 6 of Blaha (2007b), and chapter 17 of Bjorken (1965) for formulations of perturbation theory as well.

[57] We realize that such a snapshot is not possible since infinite velocity particles that could feed a camera a snapshot do not exist.

[58] In our construction of particle functional space we have not introduced complex conjugation of quantum functionals for lack of a good reason. Complex conjugation takes place only in the fourier expansion part of a

$$\mathcal{L} = \bar{\psi}_C(i\gamma^\mu D_\mu - m)\psi_C(x) + b(\bar{\psi}_C\psi_C(x))^2 \quad (11.1)$$

becomes the skeleton lagrangian

$$\mathcal{L}_S = bf^4 \quad (11.2)$$

where f is the fermion's functional.

Thus our skeletonized lagrangian formalism describes the transitions between functionals in an interaction. This formalism is made more concrete by considering Feynman diagrams for the interactions.

A skeletonized lagrangian defines a Particle Functional Theory Language. See Blaha (2005b), (2005c) and (2009)

11.2 Functional-Lagrangian and Feynman Diagrams

Feynman diagrams with their "in and out" ordering specify the transformations between functionals more completely. A simple example shows the interaction transformations of functionals. Consider the lagrangian term

$$(\bar{\psi}\psi(x))^2(\partial^\mu\varphi)^2$$

The corresponding Feynman diagram appears in Fig. 11.1.with qube functionals labeled f and quba functionals labeled b. In the above example we have not introduced internal symmetries. When internal symmetries are introduced then the skeletonized lagrangians and the corresponding Feynman diagram representations are significantly more complex.

11.3 Production Rules for Functional-Lagrangians

Each term in a Functional-Lagrangian represents a set of language production rules. See chapter 12 for an example of production rules so generated. The mapping to

---

quantum field. Another issue is the appearance of lagrangian terms with factors that are drivatives of fields. Since we do not do computations with skeleton lagrangians we can ignore the derivative in each such factor and simply substitute the functional. For example, $\varphi^3(\partial^\mu\varphi)^2$ becomes the quba expression b^5.

language production rules illustrates the language interpretation of the Unified SuperStandard Theory.

11.4 Functional Interactions and Feynman Diagrams

Feynman diagrams with their in and out ordering specify the transformations between functionals more completely. A simple example shows the interaction transformations of functionals. Consider the lagrangian term

$$(\bar{\psi}\psi(x))^2(\partial^\mu\varphi)^2$$

A corresponding Feynman diagram for it is

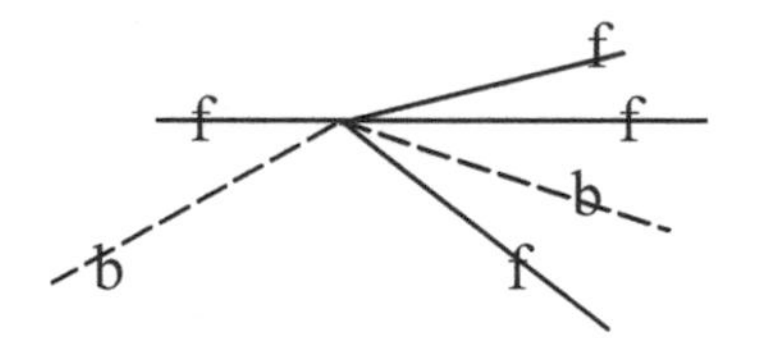

Figure 11.1.Functional Feynman diagram for the above interaction.

with qubes labeled f and qubas labeled b.

When internal symmetries are introduced, then the skeletonized lagrangians and the corresponding Feynman diagram representations are significantly more complicated.

11.5 Functional Space and Feynman Path Integrals

Functionals appear in Feynman Path Integrals and in Faddeev-Popov gauge fixing path integrals. We illustrate the use of functionals in the example:

$$Z(J) = N\int\prod_y dy \prod_\varphi d\varphi(y) \exp\{i\int d^4y[\mathscr{L}(\varphi(y)) + J^\mu(y)\varphi(y)]\} \qquad (11.3)$$

which in a functional notational notation becomes

$$Z(J) = N\int\prod_y d(y) \prod_b db \exp\{i\int d^4y[\mathscr{L}(\varphi(y)) + J^\mu(y)\varphi(y)]\} \qquad (11.4)$$

where (y) represents the fourier expansion in the y coordinates, and with the implied inner product $\varphi(y) = (b, (y))$.

12. Computational Language Interpretation of Particle Functional Transformations

In this chapter[59] we will discuss a language interpretation of particle functional transformations based on a Chomsky-like language and grammar. We will see that particle functional transformations can be viewed as grammar production rules with the net result that the evolution of the universe (Megaverse!) can be viewed as the evolution of an enormous Word consisting of a very large but finite number of (terminal) symbols.

12.1 Languages and Grammars

In chapter 3 of Blaha (2005b) we described a linguistic interpretation of particle interactions. In this interpretation particles play the role of symbols (terminal symbols and nonterminal symbols[60]) in an alphabet (of a finite number of symbols) for a Chomsky-like language. Chomsky defines four types of language ranging from type 0 through type 3. Particle theories, when viewed in terms of their perturbation expansions, can be viewed as a generalization of a type 0 language. A type 0 language (also called an unrestricted rewriting system) allows any grammar production rule of the form

$$X \rightarrow Y$$

---

[59] Most of this chapter appeared in Blaha (2005b) and other books by the author. For more than a hundred years mathematicians and physicists have been describing Physics and Mathematics as being a language in colloquial, layman's terms. Our books show that elementary particle physics, such as our SuperStandard Model, is precisely a type 0 Chomsky language, in which production rules are generated from lagrangian interaction terms. Specific examples are presented in Blaha (2005b) and (2005c).

[60] From a particle view a terminal symbol is a particle that appears in input or output states (strings) of a perturbation theory diagram. A nonterminal symbol is a particle that appears in an intermediate state of a perturbation theory diagram. In the theories that we have considered particles are both terminal and nonterminal symbols.

where X and Y are sets of particles (strings).

A production rule is a specification of a transformation of a string of symbols (set of particles) to another string of symbols (set of particles). In the case of quantum theories production rules are inherently quantum probabilistic.

A grammar is specified by a quadruple of items symbolized by the expression

<N, T, S, P>

where N is a set of nonterminal symbols, T is a set of terminal symbols, S is a special terminal symbol called the head or start symbol, and P is a finite set of production rules. In quantum theories such as the SuperStandard Theory N and T coincide. The start symbol S corresponds to the bare vacuum. Chomsky's definition of a language is the set of all strings of terminal symbols that can be generated from the start symbol using the production rules. *We extend the definition of a particle language to the set of all finite strings of particles (symbols) whether or not they can be generated from the start symbol.*[61]

The set of production rules is finite in Chomsky's definition of language. In the context of quantum field theories we note that a lagrangian of the form of a finite polynomial expression is equivalent to a finite set of production rules.

12.2 Example of Production Rules

In this section we will consider a simple example of production rules with the alphabet:

Start Symbol: S
Nonterminal symbols: A, B
Terminal symbols: x, y

We choose the production rules:

---

[61] If one considers the fact that all particles originate either directly or indirectly from the Big Bang (the Start symbol), then the Chomsky definition of type zero languages applies where all strings originate in the Start symbol of the Big Bang.

$S \rightarrow AB$	Rule I
$A \rightarrow y$	Rule II
$A \rightarrow Ay$	Rule III
$B \rightarrow x$	Rule IV
$B \rightarrow Bx$	Rule V

An example: Generating a string ('particles') yyxxx from the head symbol S using the above production rules:

$$S \rightarrow AB$$
$$AB \rightarrow AyB$$
$$AyB \rightarrow yyB$$
$$yyB \rightarrow yyBx$$
$$yyBx \rightarrow yyBxx$$
$$yyBxx \rightarrow yyxxx$$

12.3 Example of the Production Rules for a Lagrangian Interaction Term

Earlier we noted that a particle lagrangian with a finite number of terms polynomial in particle fields terms would always have a corresponding finite set of production rules. In this section we consider the example of a lagrangian electromagnetic interaction term for electrons and positrons:

$$\overline{e}\gamma\cdot Ae$$

This lagrangian term yields the production rules:

$$e \rightarrow eA$$
$$e \rightarrow Ae$$

$$eA \rightarrow e$$
$$Ae \rightarrow e$$
$$p \rightarrow pA$$
$$p \rightarrow Ap$$
$$pA \rightarrow p$$
$$Ap \rightarrow p$$
$$ep \rightarrow A$$
$$pe \rightarrow A$$
$$A \rightarrow ep$$
$$A \rightarrow pe$$

where e represents an electron, p represents a positron, and A represents the electromagnetic field. Blaha (2005b) presents sequences of transitions using the above production rules and their corresponding Feynman-like diagrams.

Blaha (2005b) also presents other examples such as the ElectroWeak Interaction:

$$\nu_e W^- e$$

where ν_e is an electron type neutrino, and W^- is a negative Weak W vector boson.

12.4 Particle Functional Transformations Identified as Production Rules

Earlier we showed how to define a lagrangian for functionals that provided transformation rules for functionals. In this section we consider the example of a functional lagrangian electromagnetic interaction term for electron and positron functionals:

$$f_e \gamma \cdot b_A f_e$$

This functional lagrangian term yields the functional production rules:

$$f_e \rightarrow f_e b_A$$
$$f_e \rightarrow b_A f_e$$
$$f_e b_A \rightarrow f_e$$
$$b_A f_e \rightarrow f_e$$
$$f_p \rightarrow f_p b_A$$
$$f_p p \rightarrow b_A f_p$$
$$f_p b_A \rightarrow f_p$$
$$b_A f_p \rightarrow f_p$$
$$f_e f_p \rightarrow b_A$$
$$f_p f_e \rightarrow b_A$$
$$b_A \rightarrow f_e f_p$$
$$b_A \rightarrow f_p f_e$$

where f_e represents an electron functional, f_p represents a positron functional, and b_A represents the electromagnetic field functional.

Space, Time, Symmetries of the Megaverse

13. Bioctonion Megaverse

In chapters 5 and 6 we developed a biquaternion theory called QUeST that could be used to derive the group structure, and the fundamental fermion and vector boson spectrums. It was based on the Unified SuperStandard Theory, where we uncovered a close similarity between the internal symmetries of the Standard Model sector and subgroups of the Lorentz group. They both exhibit U(1), SU(2) and SU(3) symmetries.

In this chapter we define a bioctonion space and use it to define a Megaverse basis of a somewhat more general Unified SuperStandard Theory, called MOST. MOST develops a more robust set of internal symmetries and creates a new view of Dark matter that appears to help explicate the lack of interactions between normal matter and Dark matter.

In chapters 5 and 6 we found the Unified SuperStandard Theory, which was based on the Complex Lorentz group, could be based on a biquaternion space for our universe. If we assume the existence of a Megaverse containing our universe, and other universes, then we can define a *Megaverse Octonion SuperStandard Theory* (MOST) that becomes a more general form of the Unified SuperStandard Theory.

13.1 Octonion Features

Octonions have significant properties that enable them to be used in a quantum field theory development:

1. An octonion is a 8-tuple of real numbers.
2. They are nonassociative.
3. They are one of the two finite dimensional division rings having the real numbers as a proper subring. (The other is quaternions—considered in chapter 5.)
4. They are non-commutative. (This is not a roadblock for quantum field theory which is also non-commutative in general.)

These features support the development of some physics theories.

13.2 Motivation and Procedure

Our goal is to create a larger dimension space within which a universe can exist and where we can derive our space-time and the Unified Superstandard Theory. Again we use the similarity of Lorentz subgroups and Standard Model internal symmetry groups to develop a deeper basis for the Unified SuperStandard Theory.

There are two possible procedures to follow in developing the deeper basis:

1. One can develop the Quantum Mechanics and Quantum Field Theory in an octonion space and then extract the dynamics, fermion spectrum, gauge fields, and so on of our familiar space-time.

2. One can define an octonion space and then use its coordinates to directly extract the space-time, internal symmetries, fermion spectrum, gauge field spectrum and dynamics.

We have chosen the latter approach as it will more directly lead to the Unified SuperStandard Theory.

In developing the deeper space, upon which we will build, we will take guidance from the derivation of the Unified SuperStandard Theory. This theory assumes a complex 4-dimensional space-time upon which Complex General Relativity is constructed. It then proceeds to complex flat space-time and Complex Relativity.

After defining features of Complex Lorentz transformations the Unified SuperStandard Theory used Lorentz boosts to derive the Dirac forms of the four fermion species. The boosts are required to boost a fermion from a rest state to a state with a real-valued energy, and real or complex-valued 3-momenta. Thus a real time–complex-valued spatial part is required for the proper definition of species.

The Unified SuperStandard Theory then showed Lorenz subgroups mapped to Standard Model internal symmetry subgroups.

13.3 Definition of Bioctonion Space

Following the above stated procedure we define a bioctonion space with *one* "time" biquarternion and *seven* "spatial" bioctonions modeled as a generalization of the 3+1 space-time of our experience. The choice of 8 bioctonion dimensions seemed natural but was not required. It does lead to an improved set of internal symmetries. We will use the symbol • to represent a bioctonion coordinate in Fig. 13.1.

We have chosen a complex 8-dimensional space-time as the Megaverse space-time. There are 128 real coordinates in the bioctonion "parent" space from which complex 8-dimensional space-time is extracted. It then eventually embeds our universe's 4-dimensional complex space-time.

Fig 13.1 symbolically depicts the space with a circle for each real-valued coordinate.

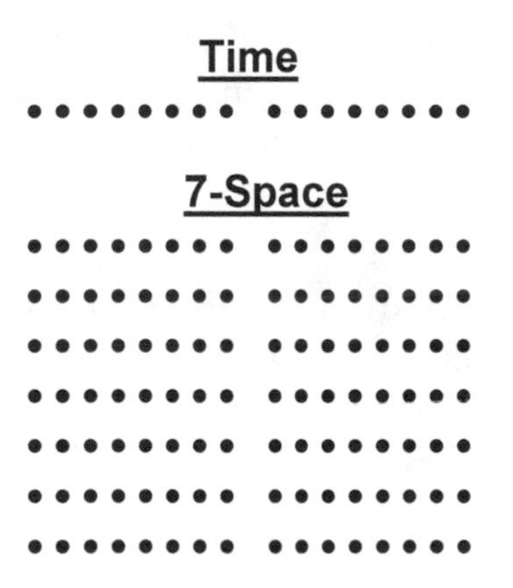

Figure 13.1. Eight-Dimensional bioctonion space with coordinates represented by • 's.

The bioctonion space has 128 real dimensions (64 complex dimensions.).

13.4 Bioctonion Lorentz Group

Our definition of time and space bioctonion coordinates purposefully resembles those of our real space-time. One might ask why there should be a Lorentz-like group for bioctonion space. The only apparent reason is the need for a special speed c in our space-time. Without c the group of the above coordinates is the fundamental representation of the bioctonion U(8) group. In this group one cannot uniquely identify a particle rest frame. The bioctonion Lorentz transformations do have a unique speed c (the speed of light) and specify a unique rest frame for any particle. Thus we must select the Bioctonionic Lorentz group for bioctonion space to have physically required, unique particle rest frames.

Flat space bioctonion Special Relativity generalizes to a bioctonion General Relativity which may be constructed directly (mindful of octonion non-commutativity).

Flat space bioctonion Lorentz group transformations have constant bioctonion matrix elements that are analogous to those of the Lorenz group. (See eq. 10.7 for a Lorentz group transformation.)

13.5 Extracting the Symmetries and Particle Spectra

As stated earlier in section 13.2 we will directly describe the symmetry structure implied by the form of the bioctonion coordinate system while mindful of sections 13.3 and 13.4.

The Unified SuperStandard Theory developed the group structure from which the particle species were derived from a subset of Lorentz boost transformations. Complex boosts mapped a system at rest to a system in motion with a real energy and complex 3-momenta in general. Bioctonion boosts play a similar role.

The 4-dimensional representation of the Unified SuperStandard Theory complex coordinates is given in Fig. 13.2.

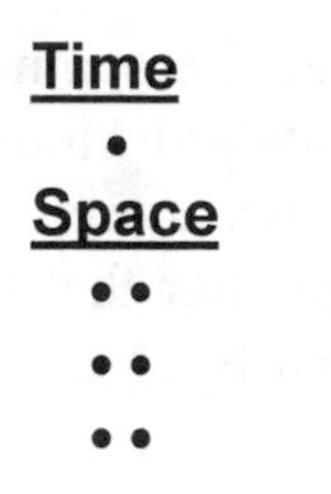

Figure 13.2. Four-Dimensional space for Unified SuperStandard derivation of particle spectra with coordinates represented by • 's.

Following the same lines we now specify a bioctonion subspace restricted to that of Fig. 13.3 to define the relevant set of coordinates for determining particle symmetries and spectra.

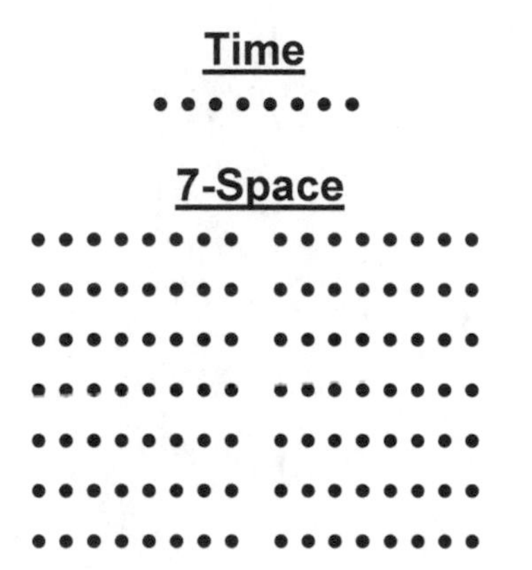

Figure 13.3. 8-dimensional bioctonion subspace for symmetries and particle spectra.

Its 60 complex coordinates will split into an 8-dimensional complex coordinate space which will support 8-dimensional Complex Lorentz transformations, and a 52 complex coordinates internal symmetry space.

The mechanism for this symmetry breakdown may be due to vacuum energy effects in the bioctonion Megaverse.

In the next chapter we analyze the symmetries of the 52 complex dimensional subspace. We find it contains the Standard Model symmetries and the Generation and Layer number symmetries. The number symmetries are inherently part of the set of internal symmetries. We will also see later that the Dark fermion sectors of the theory have spinors that occupy different parts of the overall 16 component spinors of the complex 8-dimensional space-time.

14. Symmetries of the Megaverse Octonion SuperStandard Theory (MOST)

The coordinate space picture of the Megaverse described in chapter 13 enables us to simply find the internal symmetries and particle spectra of MOST. They will turn out to be those of the Unified SuperStandard Theory as presented earlier in chapter 2.

The MOST Megaverse coordinates with the space-time complex 8-vector separated from the internal symmetry coordinates is depicted in Fig. 14.1.

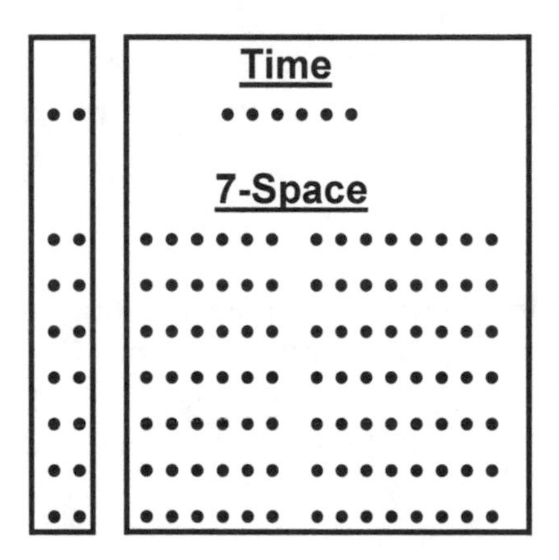

Figure 14.1. Based on Fig. 13.3; the 16 real space-time coordinates separated from 104 real coordinates for internal symmetries.

The internal symmetry coordinates above number 104 real coordinates or 52 complex coordinates. These coordinates serve as the coordinates of the fundamental representations of each of the factors of

$$[SU(2)\otimes U(1)\otimes SU(3)\otimes SU(2)\otimes U(1)\otimes SU(3)]^2\otimes U(4)^8 \tag{14.1}$$

The factorized internal symmetry emerges from another breakdown(s) which corresponds to the subgroups of the Lorentz group. They evidently follow from the structure of bioctonion Lorentz transformations.

The U(4) Generation and Layer groups are now represented in Fig. 14.1. We depict the pattern of symmetry implied by Fig. 14.1 and eq. 14.1 in Fig. 14.2.

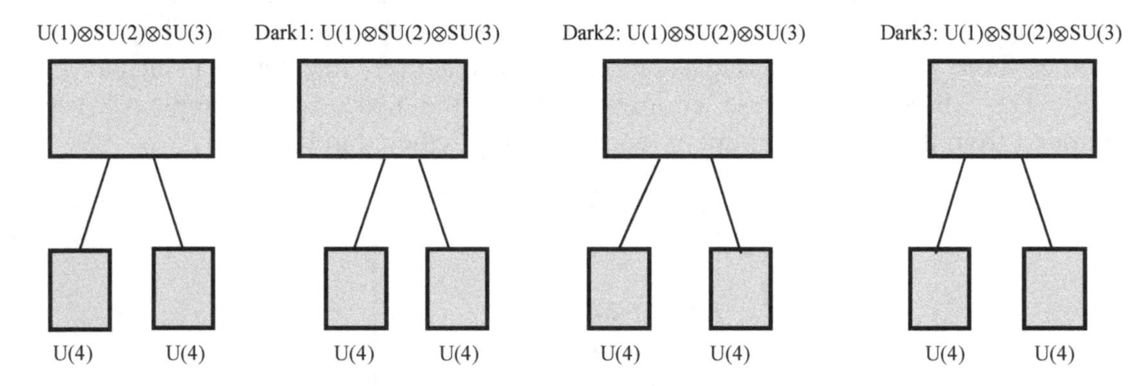

Figure 14.2. Schematic of the internal symmetry groups of eq. 14.1. The lower U(4) groups are the Generation and Layer number groups. One pair of each number group is for each of the four U(1)⊗SU(2)⊗SU(3) factors above. The result is the total Internal Symmetry group of the enlarged Unified SuperStandard Theory. (See chapter 2.)

We can separate the set of internal symmetries into four parts—three of which are Dark. The first part of the set corresponds to the known Standard Model interactions.

14.1 Layers

QUeST and MOST do not separate the vector boson interactions into four layers. Instead the layers are each flagged with a layer index. So the overall internal symmetry is the internal symmetry group of the Unified SuperStandard Theory plus two additional SU(2)⊗U(1)⊗SU(3) sectors and their Generation and Layer groups:

$$[SU(2)\otimes U(1)\otimes SU(3)\otimes SU(2)\otimes U(1)\otimes SU(3)\otimes SU(2)\otimes U(1)\otimes SU(3)\otimes SU(2)\otimes U(1)\otimes SU(3)\otimes U(4)^8]^4 \tag{14.2}$$

Whether these additional sectors exist is an experimental question. It is possible that they are only present in the Megaverse.

14.2 Fermion and Gauge Vector Boson Spectrums

The fermion and vector boson spectrums that emerge in MOST are those of the Unified SuperStandard Theory (They are displayed in Figs. 2.1 – 2.4.) plus the additional two Dark sectors. (Fig. 14.3)

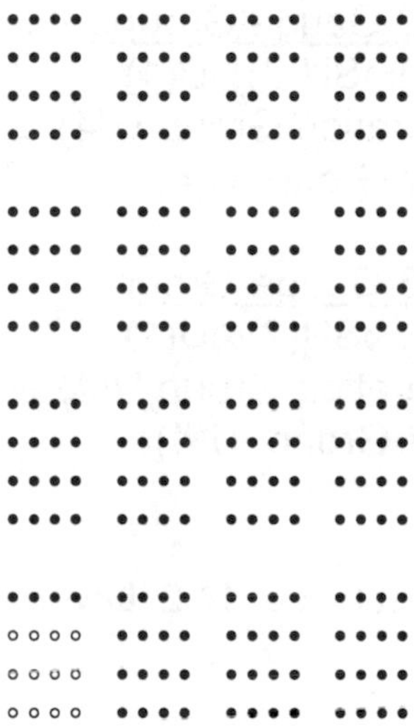

•••• •••• •••• ••••
•••• •••• •••• ••••
•••• •••• •••• ••••
•••• •••• •••• ••••

•••• •••• •••• ••••
•••• •••• •••• ••••
•••• •••• •••• ••••
•••• •••• •••• ••••

•••• •••• •••• ••••
•••• •••• •••• ••••
•••• •••• •••• ••••
•••• •••• •••• ••••

•••• •••• •••• ••••
○○○○ •••• •••• ••••
○○○○ •••• •••• ••••
○○○○ •••• •••• ••••

Figure 14.3. Schematic of the fermions of MOST. Each fermion is represented by a •. Quark triplets are represented by a single •. Four sets of four species in four generations which are in turn in 4 layers. Open symbols ○ represent known fermions. There are 512 fundamental fermions taking account of quark triplets.

<u>"Normal" Gauge Groups</u>
SU(3)⊗SU(2)⊗U(1)
Generation Group U(4)
Layer Group U(4)

<u>Dark1 Gauge Groups</u>
SU(3)⊗SU(2)⊗U(1)
Generation Group U(4)
Layer Group U(4)

<u>Dark2 Gauge Groups</u>
SU(3)⊗SU(2)⊗U(1)
Generation Group U(4)
Layer Group U(4)

<u>Dark3 Gauge Groups</u>
SU(3)⊗SU(2)⊗U(1)
Generation Group U(4)
Layer Group U(4)

Figure 14.4. MOST vector bosons list by eq. 14.2. Layers quadruple the above list.

14.3 Revised Axioms

AXIOMS

1. Bioctonion space exists and supports symmetry breaking to space-time and internal symmetries.
2. Quantum Field Theory supports fundamental particles that form a countable set. Each particle number operator is a generator in a particle interaction group.
3. All quantum field theory calculations are finite.
4. The Quantum Field Theory of particles can be defined in any curved space-time.
5. Each particle wave function has a functional defining the particle state with the functional in a set without a distance measure.

Figure 14.5 The axioms of MOST.

Now we are out of Socrates' cave.

15. Why Are Normal and Dark Particles Apparently Incapable of Interactions?

Eight-dimensional Megaverse space-time has 16 component Dirac fermion spinors. Given the split of the fermion spectrum into four sectors: normal and three dark sectors (Fig. 14.3) it is clearly reasonable to "break up" the 16-spinors into four 4-dimensional Dirac spinors for use upon entry into our 4-dimensional universe. Then each of the 16-spinors have their own 4-dimensional Dirac spinor part.

If we define all vector interactions to have a spin of the type of their own normal or Dark sector fermions (Figs. 14.2 and 14.4) then normal entities only interact with normal entities, and Dark entities of each type only interact with other Dark entities of their type.[62]

Thus we achieve a deeper justification of the absence of interactions between normal and Dark matter.

The only possible exception is the spinor interaction of gravitation whose low coupling constant value makes it negligible.

---

[62] One could go further and speculatively assert that spin has a superselection rule like electric charge so that spins of different sectors do not combine in the usual way to form total angular momentum. The superselection rule would not be possible if the gravitational spinor interaction connected the various types of fermions.

REFERENCES

Bjorken, J. D., Drell, S. D., 1964, *Relativistic Quantum Mechanics* (McGraw-Hill, New York, 1965).

Bjorken, J. D., Drell, S. D., 1965, *Relativistic Quantum Fields* (McGraw-Hill, New York, 1965).

Blaha, S., 1998, *Cosmos and Consciousness* (Pingree-Hill Publishing, Auburn, NH, 1998).

_______, 2002, *A Finite Unified Quantum Field Theory of the Elementary Particle Standard Model and Quantum Gravity Based on New Quantum Dimensions™ & a New Paradigm in the Calculus of Variations* (Pingree-Hill Publishing, Auburn, NH, 2002).

_______, 2003, *A Finite Unified Quantum Field Theory of the Elementary Particle Standard Model and Quantum Gravity Based on New Quantum Dimensions™ and a New Paradigm in the Calculus of Variations* (Pingree-Hill Publishing, Auburn, NH, 2003).

_______, 2004, *Quantum Big Bang Cosmology: Complex Space-time General Relativity, Quantum Coordinates,™ Dodecahedral Universe, Inflation, and New Spin 0, ½, 1 & 2 Tachyons & Imagyons* (Pingree-Hill Publishing, Auburn, NH, 2004).

______, *2005a, Quantum Theory of the Third Kind: A New Type of Divergence-free Quantum Field Theory Supporting a Unified Standard Model of Elementary Particles and Quantum Gravity based on a New Method in the Calculus of Variations* (Pingree-Hill Publishing, Auburn, NH, 2005).

______, 2005b, *The Metatheory of Physics Theories, and the Theory of Everything as a Quantum Computer Language* (Pingree-Hill Publishing, Auburn, NH, 2005).

______, 2005c, *The Equivalence of Elementary Particle Theories and Computer Languages: Quantum Computers, Turing Machines, Standard Model, Superstring Theory, and a Proof that Gödel's Theorem Implies Nature Must Be Quantum* (Pingree-Hill Publishing, Auburn, NH, 2005).

______, 2006a, *The Foundation of the Forces of Nature* (Pingree-Hill Publishing, Auburn, NH, 2006).

______, 2006b, *A Derivation of ElectroWeak Theory based on an Extension of Special Relativity; Black Hole Tachyons; & Tachyons of Any Spin.* (Pingree-Hill Publishing, Auburn, NH, 2006).

_______, 2007a, *Physics Beyond the Light Barrier: The Source of Parity Violation, Tachyons, and A Derivation of Standard Model Features* (Pingree-Hill Publishing, Auburn, NH, 2007).

______, 2007b, *The Origin of the Standard Model: The Genesis of Four Quark and Lepton Species, Parity Violation, the ElectroWeak Sector, Color SU(3), Three Visible Generations of Fermions, and One Generation of Dark Matter with Dark Energy* (Pingree-Hill Publishing, Auburn, NH, 2007).

______, *2008a, A Direct Derivation of the Form of the Standard Model From GL(16) (Pingree-Hill Publishing, Auburn, NH, 2008).*

______, 2008b, *A Complete Derivation of the Form of the Standard Model With a New Method to Generate Particle Masses Second Edition* (Pingree-Hill Publishing, Auburn, NH, 2008)

______, 2009, *The Algebra of Thought & Reality: The Mathematical Basis for Plato's Theory of Ideas, and Reality Extended to Include A Priori Observers and Space-Time Second Edition* (Pingree-Hill Publishing, Auburn, NH, 2009).

______, 2010a, *Operator Metaphysics: A New Metaphysics Based on a New Operator Logic and a New Quantum Operator Logic that Lead to a Mathematical Basis for Plato's Theory of Ideas and Reality* (Pingree-Hill Publishing, Auburn, NH, 2010).

______, 2010b, *The Standard Model's Form Derived from Operator Logic, Superluminal Transformations and GL(16)* (Pingree-Hill Publishing, Auburn, NH, 2010).

______, 2010c, *SuperCivilizations: Civilizations as Superorganisms* (McMann-Fisher Publishing, Auburn, NH, 2010).

______, 2011a, *21st Century Natural Philosophy Of Ultimate Physical Reality* (McMann-Fisher Publishing, Auburn, NH, 2011).

______, 2011b, *All the Universe! Faster Than Light Tachyon Quark Starships & Particle Accelerators with the LHC as a Prototype Starship Drive Scientific Edition* (Pingree-Hill Publishing, Auburn, NH, 2011).

______, 2011c, *From Asynchronous Logic to The Standard Model to Superflight to the Stars* (Blaha Research, Auburn, NH, 2011).

______, 2012a, *From Asynchronous Logic to The Standard Model to Superflight to the Stars volume 2: Superluminal CP and CPT, U(4) Complex General Relativity and The Standard Model, Complex Vierbein General Relativity, Kinetic Theory, Thermodynamics* (Blaha Research, Auburn, NH, 2012).

______, 2012b, *Standard Model Symmetries, And Four And Sixteen Dimension Complex Relativity; The Origin Of Higgs Mass Terms* (Blaha Reasearch, Auburn, NH, 2012).

______, 2013a, *Multi-Stage Space Guns, Micro-Pulse Nuclear Rockets, and Faster-Than-Light Quark-Gluon Ion Drive Starships* (Blaha Research, Auburn, NH, 2013).

______, 2013b, *The Bridge to Dark Matter; A New Sister Universe; Dark Energy; Inflatons; Quantum Big Bang; Superluminal Physics; An Extended Standard Model Based on Geometry* (Blaha Reasearch, Auburn, NH, 2013).

______, 2014a, *Universes and Megaverses: From a New Standard Model to a Physical Megaverse; The Big Bang; Our Sister Universe's Wormhole; Origin of the Cosmological Constant, Spatial Asymmetry of the Universe, and its Web of Galaxies; A Baryonic Field between Universes and Particles; Megaverse Extended Wheeler-DeWitt Equation* (Blaha Reasearch, Auburn, NH, 2014).

______, 2014b, *All the Megaverse! Starships Exploring the Endless Universes of the Cosmos Using the Baryonic Force* (Blaha Research, Auburn, NH, 2014).

______, 2014c, *All the Megaverse! II Between Megaverse Universes: Quantum Entanglement Explained by the Megaverse Coherent Baryonic Radiation Devices – PHASERs Neutron Star Megaverse Slingshot Dynamics Spiritual and UFO Events, and the Megaverse Microscopic Entry into the Megaverse* (Blaha Research, Auburn, NH, 2014).

______, 2015a, *PHYSICS IS LOGIC PAINTED ON THE VOID: Origin of Bare Masses and The Standard Model in Logic, U(4) Origin of the Generations, Normal and Dark Baryonic Forces, Dark Matter, Dark Energy, The Big Bang, Complex General Relativity, A Megaverse of Universe Particles* (Blaha Research, Auburn, NH, 2015).

______, 2015b, *PHYSICS IS LOGIC Part II: The Theory of Everything, The Megaverse Theory of Everything, U(4)⊗U(4) Grand Unified Theory (GUT), Inertial Mass = Gravitational Mass, Unified Extended Standard Model and a New Complex General Relativity with Higgs Particles, Generation Group Higgs Particles* (Blaha Research, Auburn, NH, 2015).

______, 2015c, *The Origin of Higgs ("God") Particles and the Higgs Mechanism: Physics is Logic III, Beyond Higgs – A Revamped Theory With a Local Arrow of Time, The Theory of Everything Enhanced, Why Inertial Frames are Special, Universes of the Mind* (Blaha Research, Auburn, NH, 2015).

______, 2015d, *The Origin of the Eight Coupling Constants of The Theory of Everything: U(8) Grand Unified Theory of Everything (GUTE),* S^8 *Coupling Constant Symmetry, Space-Time Dependent Coupling Constants, Big Bang Vacuum Coupling Constants, Physics is Logic IV* (Blaha Research, Auburn, NH, 2015).

______, 2016a, *New Types of Dark Matter, Big Bang Equipartition, and A New U(4) Symmetry in the Theory of Everything: Equipartition Principle for Fermions, Matter is 83.33% Dark,*

Penetrating the Veil of the Big Bang, Explicit QFT Quark Confinement and Charmonium, Physics is Logic V (Blaha Research, Auburn, NH, 2016).

______, 2016b, *The Periodic Table of the 192 Quarks and Leptons in The Theory of Everything: The U(4) Layer Group, Physics is Logic VI* (Blaha Research, Auburn, NH, 2016).

______, 2016c, *New Boson Quantum Field Theory, Dark Matter Dynamics, Dark Matter Fermion Layer Mixing, Genesis of Higgs Particles, New Layer Higgs Masses, Higgs Coupling Constants, Non-Abelian Higgs Gauge Fields, Physics is Logic VII* (Blaha Research, Auburn, NH, 2016).

______, 2016d, *Unification of the Strong Interactions and Gravitation: Quark Confinement Linked to Modified Short-Distance Gravity; Physics is Logic VIII* (Blaha Research, Auburn, NH, 2016).

______, 2016e, *MoND: Unification of the Strong Interactions and Gravitation II, Quark Confinement Linked to Large-Scale Gravity, Physics is Logic IX* (Blaha Research, Auburn, NH, 2016).

______, 2016f, *CQ Mechanics: A Unification of Quantum & Classical Mechanics, Quantum/Semi-Classical Entanglement, Quantum/Classical Path Integrals, Quantum/Classical Chaos* (Blaha Research, Auburn, NH, 2016).

______, 2016g, *GEMS: Unified Gravity, ElectroMagnetic and Strong Interactions: Manifest Quark Confinement, A Solution for the Proton Spin Puzzle, Modified Gravity on the Galactic Scale* (Pingree Hill Publishing, Auburn, NH, 2016).

______, 2016h, *Unification of the Seven Boson Interactions based on the Riemann-Christoffel Curvature Tensor* (Pingree Hill Publishing, Auburn, NH, 2016).

______, 2017a, *Unification of the Eleven Boson Interactions based on 'Rotations of Interactions'* (Pingree Hill Publishing, Auburn, NH, 2017).

______, 2017b, *The Origin of Fermions and Bosons, and Their Unification* (Pingree Hill Publishing, Auburn, NH, 2017).

______, 2017c, *Megaverse: The Universe of Universes* (Pingree Hill Publishing, Auburn, NH, 2017).

______, 2017d, *SuperSymmetry and the Unified SuperStandard Model* (Pingree Hill Publishing, Auburn, NH, 2017).

______, 2017e, *From Qubits to the Unified SuperStandard Model with Embedded SuperStrings: A Derivation* (Pingree Hill Publishing, Auburn, NH, 2017).

______, 2017f, *The Unified SuperStandard Model in Our Universe and the Megaverse: Quarks, ...* , (Pingree Hill Publishing, Auburn, NH, 2017).

______, 2018a, *The Unified SuperStandard Model and the Megaverse SECOND EDITION A Deeper Theory based on a New Particle Functional Space that Explicates Quantum Entanglement Spookiness (Volume 1)* (Pingree Hill Publishing, Auburn, NH, 2018).

______, 2018b, *Cosmos Creation: The Unified SuperStandard Model, Volume 2, SECOND EDITION* (Pingree Hill Publishing, Auburn, NH, 2018).

______, 2018c, *God Theory (*Pingree Hill Publishing, Auburn, NH, 2018).

______, 2018d, *Immortal Eye: God Theory: Second Edition* (Pingree Hill Publishing, Auburn, NH, 2018).

______, 2018e, *Unification of God Theory and Unified SuperStandard Model THIRD EDITION* (Pingree Hill Publishing, Auburn, NH, 2018).

______, 2019a, *Calculation of: QED α = 1/137, and Other Coupling Constants of the Unified SuperStandard Theory* (Pingree Hill Publishing, Auburn, NH, 2019).

______, 2019b, *Coupling Constants of theUnified SuperStandard Theory SECOND EDITION* (Pingree Hill Publishing, Auburn, NH, 2019).

______, 2019c, *New Hybrid Quantum Big_Bang–Megaverse_Driven Universewith a Finite Big Bang and an Increasing Hubble Constant* (Pingree Hill Publishing, Auburn, NH, 2019).

______, 2019d, *The Universe, The Electron and The Vacuum* (Pingree Hill Publishing, Auburn, NH, 2019).

______, 2019e, *Quantum Big Bang – Quantum Vacuum Universes (Particles)* (Pingree Hill Publishing, Auburn, NH, 2019).

______, 2019f, *The Exact QED Calculation of the Fine Structure Constant Implies ALL 4D Universes have the Same Physics/Life Prospects* (Pingree Hill Publishing, Auburn, NH, 2019).

______, 2019g, *Unified SuperStandard Theory and the SuperUniverse Model: The Foundation of Science* (Pingree Hill Publishing, Auburn, NH, 2018).

Eddington, A. S., 1952, *The Mathematical Theory of Relativity* (Cambridge University Press, Cambridge, U.K., 1952).

Fant, Karl M., 2005, *Logically Determined Design: Clockless System Design With NULL Convention Logic* (John Wiley and Sons, Hoboken, NJ, 2005).

Feinberg, G. and Shapiro, R., 1980, *Life Beyond Earth: The Intelligent Earthlings Guide to Life in the Universe* (William Morrow and Company, New York, 1980).

Gelfand, I. M., Fomin, S. V., Silverman, R. A. (tr), 2000, *Calculus of Variations* (Dover Publications, Mineola, NY, 2000).

Giaquinta, M., Modica, G., Souchek, J., 1998, *Cartesian Coordinates in the Calculus of Variations* Volumes I and II (Springer-Verlag, New York, 1998).

Giaquinta, M., Hildebrandt, S., 1996, *Calculus of Variations* Volumes I and II (Springer-Verlag, New York, 1996).

Gradshteyn, I. S. and Ryzhik, I. M., 1965, *Table of Integrals, Series, and Products* (Academic Press, New York, 1965).

Heitler, W., 1954, *The Quantum Theory of Radiation* (Claendon Press, Oxford, UK, 1954).

Huang, Kerson, 1992, *Quarks, Leptons & Gauge Fields 2nd Edition* (World Scientific Publishing Company, Singapore, 1992).

Jost, J., Li-Jost, X., 1998, *Calculus of Variations* (Cambridge University Press, New York, 1998).

Kaku, Michio, 1993, *Quantum Field Theory*, (Oxford University Press, New York, 1993).

Kirk, G. S. and Raven, J. E., 1962, *The Presocratic Philosophers* (Cambridge University Press, New York, 1962).

Landau, L. D. and Lifshitz, E. M., 1987, *Fluid Mechanics 2nd Edition*, (Pergamon Press, Elmsford, NY, 1987).

Misner, C. W., Thorne, K. S., and Wheeler, J. A., 1973, *Gravitation* (W. H. Freeman, New York, 1973).

Rescher, N., 1967, *The Philosophy of Leibniz* (Prentice-Hall, Englewood Cliffs, NJ, 1967).

Rieffel, Eleanor and Polak, Wolfgang, 2014, *Quantum Computing* (MIT Press, Cambridge, MA, 2014).

Riesz, Frigyes and Sz.-Nagy, Béla, 1990, *Functional Analysis* (Dover Publications, New York, 1990).

Sagan, H., 1993, *Introduction to the Calculus of Variations* (Dover Publications, Mineola, NY, 1993).

Sakurai, J. J., 1964, *Invariance Principles and Elementary Particles* (Princeton University Press, Princeton, NJ, 1964).

Sorokin, Pitirim, 1941, *Social and Cultural Dynamics* (Porter Sargent Publishers, Boston, MA, 1941).

Streater, R. F. and Wightman, A. S., 2000, *PCT, Spin, Statistics, and All That* (Princeton University Press, Princeton, NJ 2000).

Weinberg, S., 1972, *Gravitation and Cosmology* (John Wiley and Sons, New York, 1972).

Weinberg, S., 1995, *The Quantum Theory of Fields Volume I* (Cambridge University Press, New York, 1995).

Weinberg, S., 2000, *The Quantum Theory of Fields Volume III Supersymmetry* (Cambridge University Press, New York, 2000).

Weyl, H., 1950, *Space, Time, Matter* (Dover, New York, 1950).

Weyl, H., (Tr. S. Pollard et al), 1987, *The Continuum* (Dover Publications, New York, 1987).

INDEX

About the Author

Stephen Blaha is a well-known Physicist and Man of Letters with interests in Science, Society and civilization, the Arts, and Technology. He had an Alfred P. Sloan Foundation scholarship in college. He received his Ph.D. in Physics from Rockefeller University. He has served on the faculties of several major universities. He was also a Member of the Technical Staff at Bell Laboratories, a manager at the Boston Globe Newspaper, a Director at Wang Laboratories, and President of Blaha Software Inc. and of Janus Associates Inc. (NH).

Among other achievements he was a co-discoverer of the "r potential" for heavy quark binding developing the first (and still the only demonstrable) non-abelian gauge theory with an "r" potential; first suggested the existence of topological structures in superfluid He-3; first proposed Yang-Mills theories would appear in condensed matter phenomena with non-scalar order parameters; first developed a grammar-based formalism for quantum computers and applied it to elementary particle theories; first developed a new form of quantum field theory without divergences (thus solving a major 60 year old

problem that enabled a unified theory of the Standard Model and Quantum Gravity without divergences to be developed); first developed a formulation of complex General Relativity based on analytic continuation from real space-time; first developed a generalized non-homogeneous Robertson-Walker metric that enabled a quantum theory of the Big Bang to be developed without singularities at t = 0; first generalized Cauchy's theorem and Gauss' theorem to complex, curved multi-dimensional spaces; received Honorable Mention in the Gravity Research Foundation Essay Competition in 1978; first developed a physically acceptable theory of faster-than-light particles; first derived a composition of extrema method in the Calculus of Variations; first quantitatively suggested that inflationary periods in the history of the universe were not needed; first proved Gödel's Theorem implies Nature must be quantum; provided a new alternative to the Higgs Mechanism, and Higgs particles, to generate masses; first showed how to resolve logical paradoxes including Gödel's Undecidability Theorem by developing Operator Logic and Quantum Operator Logic; first developed a quantitative harmonic oscillator-like model of the life cycle, and interactions, of civilizations; first showed how equations describing superorganisms also apply to civilizations. A recent book shows his theory applies successfully to the past 14 years of history and to *new* archaeological data on Andean and Mayan civilizations as well as Early Anatolian and Egyptian civilizations.

He first developed an axiomatic derivation of the form of The Standard Model from geometry – space-time properties – The Unified SuperStandard Theory. It unifies all the known forces of Nature. It also has a Dark Matter sector that includes a Dark ElectroWeak sector with Dark doublets and Dark gauge interactions. It uses quantum coordinates to remove infinities that crop up in most interacting quantum field theories and additionally to remove the infinities that appear in the Big Bang and generate inflationary growth of the universe. It shows gravity has a MOND-like form without sacrificing Newton's Laws. It relates the interactions of the MOND-like sector of gravity with the r-potential of Quark Confinement. The axioms of the theory lead to the question of their origin. We suggest in the preceding edition of this book it can be attributed to an

entity with God-like properties. We explore these properties in "God Theory" and show they predict that the Cosmos exists forever although individual universes (or incarnations of our universe) "come and go." Several other important results emerge from God Theory such a functionally triune God. The Unified SuperStandard Theory has many other important parts described in the Current Edition of *The Unified SuperStandard Theory* and expanded in subsequent volumes.

Blaha has had a major impact on a succession of elementary particle theories: his Ph.D. thesis (1970), and papers, showed that quantum field theory calculations to all orders in ladder approximations could not give scaling deep inelastic electron-nucleon scattering. He later showed the eigenvalue equation for the fine structure constant α in Johnson-Baker-Willey QED had a zero at α = 1 not 1/137 by solving the Schwinger-Dyson equations to all orders in an approximation that agreed with exact results to 4th order in α thus ending interest in this theory. In 1979 at Prof. Ken Johnson's (MIT) suggestion he calculated the proton-neutron mass difference in the MIT bag model and found the result had the wrong sign reducing interest in the bag model. These results all appear in Physical Review papers. In the 2000's he repeatedly pointed out the shortcomings of SuperString theory and showed that The Standard Model's form could be derived from space-time geometry by an extension of Lorentz transformations to faster than light transformations. This deeper space-time basis greatly increases the possibility that it is part of THE fundamental theory. Recently, Blaha showed that the Weak interactions differed significantly from the Strong, electromagnetic and gravitation interactions in important respects while these interactions had similar features, and suggested that ElectroWeak theory, which is essentially a glued union of the Weak interactions and Electromagnetism, possibly modulo unknown Higgs particle features, be replaced by a unified theory of the other interactions combined with a stand-alone Weak interaction theory. Blaha also showed that, if Charmonium calculations are taken seriously, the Strong interaction coupling constant is only a factor of five larger than the electromagnetic coupling constant, and thus Strong interaction perturbation theory would make sense and yield physically meaningful results.

In graduate school (1965-71) he wrote substantial papers in elementary particles and group theory: The Inelastic E- P Structure Functions in a Gluon Model. Phys. Lett. B40:501-502,1972; Deep-Inelastic E-P Structure Functions In A Ladder Model With Spin 1/2 Nucleons, Phys.Rev. D3:510-523,1971; Continuum Contributions To The Pion Radius, Phys. Rev. 178:2167-2169,1969; Character Analysis of U(N) and SU(N), J. Math. Phys. 10, 2156 (1969); and The Calculation of the Irreducible Characters of the Symmetric Group in Terms of the Compound Characters, (Published as Blaha's Lemma in D. E. Knuth's book: *The Art of Computer Programming Vols. 1 – 4*).

In the early 1980's Blaha was also a pioneer in the development of UNIX for financial, scientific and Internet applications: benchmarked UNIX versions showing that block size was critical for UNIX performance, developing financial modeling software, starting database benchmarking comparison studies, developing Internet-like UNIX networking (1982) and developing a hybrid shell programming technique (1982) that was a precursor to the PERL programming language. He was also the manager of the AT&T ten-year future products development database. His work helped lead to commercial UNIX on computers such as Sun Micros, IBM AIX minis, and Apple computers.

In the 1980's he pioneered the development of PC Desktop Publishing on laser printers. and was nominated for three "Awards for Technical Excellence" in 1987 by PC Magazine for PC software products that he designed and developed.

He has developed a theory of Megaverses – actual universes of which our universe is one – with quantum particle-like properties based on the Wheeler-DeWitt equation of Quantum Gravity. He has developed a theory of a baryonic force, which had been conjectured many years ago, and estimated the strength of the force based on discrepancies in measurements of the gravitational constant G. This force, operative in D-dimensional space, can be used to escape from our

universe in “uniships” which are the equivalent of the faster-than-light starships proposed in the author’s earlier books. Thus travel to other universes, as well as to other stars is possible.
Blaha also considered the complexified Wheeler-DeWitt equation and showed that its limitation to real-valued coordinates and metrics generated a Cosmological Constant in the Einstein equations.
Recently he calculated the QED Fine Structure Constant exactly to the experimentally known 13 places. He also used the same approach to approximately calculate the Weak interaction SU(2) and Strong interaction SU(3) coupling constants successfully. Based on the origin of all coupling constants in quantum field theoretic vacuum polarization effects he suggested all universes would have the same interactions and consequently the same Physics, Chemistry and Biology. Thus all universes would be trivially Anthropic and capable of Life. Going further he suggested universes are particles and that universe expansion is co9mpletely analogous to vacuum polarization of a universe particle due to a gauge vector universe interaction. Universe expansion as a function of time is the fourier transform of universe vacuum polarization. This feature was demonstrated by almost exact (when compared to the universe scale factor) calculation of the small times universe scale factor in perturbation theory.
The author has also recently written a series of books on the serious problems of the United States and their solution as well as a book on the decline of Mankind that will follow from current social and genetic trends in Mankind.

In the past twelve years Dr. Blaha has written over 40 books on a wide range of topics. Some recent major works are: *From Asynchronous Logic to The Standard Model to Superflight to the Stars*, *All the Universe!*, *SuperCivilizations: Civilizations as Superorganisms*, *America's Future: an Islamic Surge, ISIS, al Qaeda, World Epidemics, Ukraine, Russia-China Pact, US Leadership Crisis*, *The Rises and Falls of Man – Destiny – 3000 AD: New Support for a Superorganism MACRO-THEORY of CIVILIZATIONS From CURRENT WORLD TRENDS and NEW Peruvian, Pre-Mayan, Mayan, Anatolian, and Early Egyptian Data, with a Projection to 3000 AD*, and *Mankind in Decline: Genetic Disasters, Human-Animal Hybrids, Overpopulation, Pollution, Global Warming, Food and Water Shortages, Desertification, Poverty, Rising Violence, Genocide, Epidemics, Wars, Leadership Failure.*

He has taught approximately 4,000 students in undergraduate, graduate, and postgraduate corporate education courses primarily in major universities, and large companies and government agencies.